AA

SPOTTER GUIDE



# BUTTERFLIES & MOTHS

Written by Andrew Cleave

Produced for AA Publishing by D & N Publishing, Baydon, Wiltshire
Commissioning editor at AA Publishing: Paul Mitchell
Production at AA Publishing: Rachel Davis

Printed and bound in China by C&C Offset Printing Co. Ltd

A CIP catalogue record for this book is available from the British Library.

ISBN 978 0 7495 68290
978 0 7495 68450 (SS)



Published by AA Publishing, a trading name of AA Media Limited, whose registered office is Fanum House, Basing View, Basingstoke, Hampshire RG21 4EA. Registered number 06112600.

A04089
theAA.com/shop

# CONTENTS

# INTRODUCTION

BUTTERFLIES AND MOTHS are arguably the most popular insects in Britain, and few people have a bad word to say about the former in particular, apart perhaps from Large Whites in the garden, whose larvae eat our cabbages. Most butterflies are extremely colourful and well marked, and are active in the daytime. Although some moth species are diurnal, most fly by night and their wing markings often provide camouflage from prying predatory eyes when they are resting in the daytime.

The *AA Spotter Guide to Butterflies and Moths* covers 58 species of butterflies – all the species anyone is likely to see in England, Wales and Scotland; many of these butterflies are common and widespread, but some are local or have very specific habitat requirements. In addition, 124 species of moths have been included in the book. Although several thousand moth species have been recorded in Britain, those described here represent Britain's most common and easily discovered moths, plus a few stunning and unusual species to whet your appetite.

***SILVER-STUDDED BLUE,* one of our most beautiful butterflies.**

A full page is devoted to each species, and each one begins with the common English name (where one exists) and is followed by the species' scientific name.

For ease of use, the subsequent text has been divided into sections: **FACT FILE**, which covers the species' size, habitat preferences, larval foodplant and adult flight period; **IDENTIFICATION**, which describes its appearance; **STATUS AND COMMENTS**, which describes where the species occurs in the region, and provides an indication of its abundance or scarcity; and **KEY FACT**, which gives tips to help the reader find or identify the species in question, or distinguish it from a close relative. A stunning photograph accompanies the text.

# FINDING AND STUDYING BUTTERFLIES AND MOTHS



BUTTERFLIES ARE EASILY observed when the sun is shining, and if you have a colourful display of flowers in your garden you can usually record a dozen or more species there in midsummer. If you grow nectar-rich plants such as Butterfly-bush, Hebe or Sedum, you will attract an even greater diversity and number of butterflies. Don't forget to include their larval foodplants too (read the text for advice), because without these the butterfly life cycle cannot be completed.

Mature woodlands have their own range of special butterflies (fritillary species, White Admirals and Speckled Woods among them), and sunny woodland rides in particular can be very rewarding, especially when Bramble (a good nectar source) is in flower. Our heathlands are favoured by specialities such as Grayling and Silver-studded Blue, while good-quality, unspoilt grassland usually supports several skipper species, plus Meadow Browns, Marbled Whites and Gatekeepers. Chalk grassland is a particularly special habitat and should be visited to see Chalkhill and Adonis blues, and Silver-spotted Skippers. Wherever you visit, if it is a dull day butterflies may sometimes be resting on plants, allowing close-up views to be obtained.

Moths can be found resting in the day. Look for them on tree bark, on wooden fence panels or amongst shady vegetation. At dusk, a few moth species can be seen feeding in flower borders in the garden, but generally the greatest numbers are discovered when they are attracted to light after dark. People who are particularly keen on studying moths use a light source that emits ultraviolet light (moth eyes are particularly sensitive to this) to attract them and a box trap to keep them captive overnight (they are then released unharmed). But any reasonably bright bulb will attract some species, especially if it is placed against a backdrop of a white sheet or wall.

Apart from a few success stories, butterflies and moths – like much of our wildlife – are under threat and many species are in decline. Habitat destruction by farmers, habitat loss to developers and inappropriate site management all play their part in these insects' plight. If you want to encourage these wonderful creatures, support conservation bodies, in particular the organisation Butterfly Conservation – visit its website (www.butterfly-conservation.org) for more information. And it goes without saying that no butterflies or moths should be collected; instead, record your sightings and discoveries with a camera!

# WHAT ARE BUTTERFLIES AND MOTHS?

BUTTERFLIES AND MOTHS are insects, AND are related to invertebrates such as beetles, bugs, caddisflies and grasshoppers. Like their relatives, adult butterflies and moths have three pairs of legs and a body that is divided into three sections: the head, on which the mouthparts, eyes and sensory antennae are located; the thorax, to which the legs and two pairs of wings are attached; and the segmented abdomen, inside which many vital organs are found.

All butterflies and moths have four stages in their life cycle: adults mate and females lay eggs; small larvae (sometimes called caterpillars) hatch from the eggs and, through a succession of moults, increase in size; when fully grown, the larvae pupate; and dramatic changes within the pupa (sometimes called a chrysalis) create the adult insect, which emerges by splitting open the pupal case. These different stages can sometimes all be found in the garden. If you find a larva, and if you are certain of the foodplant on which it is feeding, it can be fun to watch it grow: place it in an airy container and provide fresh leaves on a daily basis; after a few weeks it should pupate, and the adult will emerge either later in the year or the following spring.

***RED ADMIRAL BUTTERFLY***

**Abdomen** Hind section of a butterfly or moth's body; usually appears segmented.

**Antennae** Slender, paired sensory organs on a butterfly or moth head.

**Caterpillar** Growing stage in the life cycle of a butterfly or moth; usually referred to as a larva.

**Compound eye** Eye structure typical of butterflies and moths, and insects generally, comprising numerous cells and lenses.

**Diurnal** Flying in the daytime; all butterflies, and a few moth species, are diurnal.

**Downland** Habitat of grazed, herb-rich grassland on chalky soils; the best butterfly downland is in Hampshire, Wiltshire, Sussex and Kent.

**Head** Front section of a butterfly or moth's body, where the eyes, antennae and mouthparts are located.

**Heathland** Habitat found on acid, sandy soils and dominated by plants such as Heather, Bell Heather and gorse species; the best heathland habitats are in Hampshire, Dorset and Surrey.

**Instar** Stage between moults in the life of a butterfly or moth larva.

**Larva** Soft-bodied, growing stage in the life cycle of butterflies and moths; also known as a caterpillar.

**Metamorphosis** The changes that take place in the life cycle of a butterfly or moth – for example, when a larva turns into a pupa, or an adult insect emerges from a pupa.

**Moult** Shedding the skin as part of the growing process in butterfly and moth larvae.

**Nectar** Sugar- and nutrient-rich fluid found in flowers.

**Nocturnal** Flying after dark; not all moths fly at night but butterflies are never nocturnal.

**Palps** Sensory appendages found around the mouth of adult butterflies and moths.

**Proboscis** Long, extensible, mouth-part, designed for sucking nectar or other fluids; held coiled at rest.

**Pupa** Resting stage in the life cycle of a butterfly or moth, and from which the adult emerges; sometimes referred to as a chrysalis.

**Pupation** The process whereby a butterfly or moth larva metamorphoses into a pupa.

**Thorax** The middle section of a moth or butterfly's body, to which the wings and legs are attached.

**Wingspan** Distance from one wingtip to the other in a butterfly or moth.

# SWALLOWTAIL

*Papilio machaon ssp. britannicus*

FACT FILE

**SIZE** Wingspan 70mm **HABITAT** Fens and marshes

**LARVAL FOODPLANT** Milk-parsley

**FLIGHT PERIOD** Two broods: May–Jun and Aug

## IDENTIFICATION

Large, unmistakable butterfly. Upper forewings have yellow and black markings; upper hindwings are yellow and black, with a blue suffusion on outer dark band, a red and blue eyespot, and a long tail streamer. Underwing pattern is similar to that on upperwings, with yellow predominating, and more extensive red on hindwing.

## STATUS AND COMMENTS

Extremely local and confined to a few East Anglian fens, mainly in the Norfolk Broads. Land management favouring the Swallowtail's larval foodplant is essential for the species' survival.

## KEY FACT

Spring broods of Swallowtails can sometimes be seen feeding on Ragged-robin flowers. Summer broods will sometimes visit Butterfly-bushes in gardens near their marshland habitats.

## SPOTTER'S CHART

| LOCATION | DATE/TIME |
|---|---|
| | |
| | |
| | |
| | |
| | |

# WOOD WHITE

*Leptidea sinapis*

## FACT FILE

**SIZE** Wingspan 40mm **HABITAT** Open woodland, rides and scrub **LARVAL FOODPLANT** Pea family members, including Bitter-vetch **FLIGHT PERIOD** Two broods: May–Jun and Jul–Aug

## IDENTIFICATION

Delicate little butterfly with rounded wings. Upperwings are off-white and faintly marbled with grey on hindwing; forewing has a dark tip, more intense in male than female. Underwings show a similar pattern but grey marbling is more intense while dark forewing tip is paler.

## STATUS AND COMMENTS

Extremely local, with its main colonies in SW and central England, and along the Welsh border. Seldom numerous; in some years it is scarce.

## KEY FACT

The Wood White's flight is weak and fluttery at best; on dull days, or when the sun is hidden by a cloud, it drops into the cover of vegetation and rests. It always settles with its wings closed.

## SPOTTER'S CHART

| LOCATION | DATE/TIME |
|---|---|
| | |
| | |
| | |
| | |
| | |
| | |

# LARGE WHITE

*Pieris brassicae*

**FACT FILE**

**SIZE** Wingspan 60mm **HABITAT** Gardens, parks and open countryside **LARVAL FOODPLANT** Cabbage, other brassicas and Nasturtium **FLIGHT PERIOD** Two broods: May–Jun and Aug–Sep

## IDENTIFICATION

Britain's largest 'white' butterfly. Upperwings are creamy white with a black tip to forewing and a dark spot on leading edge of hindwing; female also has two black spots on forewing. Underside of hindwing is yellow; forewing is white with a yellow tip, and with dark spots in female.

## STATUS AND COMMENTS

Common and widespread but less numerous than in the past. Often seen in gardens, where the yellow and black larvae strip the leaves of cultivated Cabbages; they hatch from yellow eggs, laid in clusters.

**KEY FACT** The Large White spends the winter as a pupa, and adults emerge in spring. Numbers are also boosted during the summer months by annual immigration from mainland Europe.

## SPOTTER'S CHART

| LOCATION | DATE/TIME |
| --- | --- |
| | |
| | |
| | |
| | |
| | |

# SMALL WHITE

*Pieris rapae*

**FACT FILE** SIZE Wingspan 45mm HABITAT Gardens, parks and open countryside LARVAL FOODPLANT Cabbage, other brassicas and Nasturtium FLIGHT PERIOD Two broods: Apr–May and Jul–Aug

## IDENTIFICATION

Smaller than Large White. Upperwings are creamy white with a dark grey tip; female has two black spots on forewing and one on leading edge of hindwing; on male, spots are extremely pale. Underside of hindwing is yellow; forewing is white with a yellow tip and faint pattern of upperwing's dark spots.

## STATUS AND COMMENTS

Common and widespread, except in the far N. Often seen in gardens, where the larvae (green with yellow spots) strip Cabbage leaves; the yellow eggs are laid singly.

## KEY FACT

The Small White spends the winter months as a pupa, which is either attached to a plant stem or a fence. Numbers are boosted in summer by immigration from mainland Europe.

## SPOTTER'S CHART

| LOCATION | DATE/TIME |
| --- | --- |
| | |
| | |
| | |
| | |

# GREEN-VEINED WHITE

*Pieris napi*

**FACT FILE**

**SIZE** Wingspan 45–50mm **HABITAT** Hedgerows, woods and gardens **LARVAL FOODPLANT** Garlic and Hedge mustard, and Cuckooflower **FLIGHT PERIOD** Two broods: Apr–Jun and Jul–Aug

## IDENTIFICATION

Upperwings are white with grey veins that are darker in female than male. Forewing has a grey tip and in female also has two dark spots. Underside of hindwing is yellow with dark grey-green veins; forewing is pale with dark veins, and in female also has two dark spots.

## STATUS AND COMMENTS

Common and widespread, least so in N and upland Britain. Most likely seen flying along country hedgerows and lanes, although it is also an occasional visitor to gardens.

**KEY FACT**

The vein markings – smoky grey on the upperwings and greenish grey on the underwings – vary in intensity between individuals but are always sufficiently obvious and striking to separate this species from the similar Small White.

## SPOTTER'S CHART

| LOCATION | DATE/TIME |
|---|---|
| | |
| | |
| | |

# ORANGE-TIP

*Anthocharis cardamines*

**FACT FILE** SIZE Wingspan 40mm HABITAT Hedgerows, woodland rides and rural gardens LARVAL FOODPLANT Cuckooflower, Garlic Mustard and other crucifers FLIGHT PERIOD Apr–Jun

## IDENTIFICATION

Well-marked butterfly with rounded wings. Male has bright orange outer half to forewing; upperwings are otherwise white but subtly marbled with grey on hindwing. Underwing pattern and colours reflect those on upperwing, but hindwing marbling is an intense greenish grey. Female is similar to male but lacks any orange.

## STATUS AND COMMENTS

Fairly common and widespread, least so in N and upland Britain. Often seen flying through rural gardens, and a common sight in spring woodlands and hedgerows.

**KEY FACT** A male Orange-tip is unmistakable but, in flight, a female could be confused with a Green-veined White; at rest, note the female Orange-tip's marbled, not dark-veined, hindwings. Orange-tip larvae are cannibalistic.

## SPOTTER'S CHART

| LOCATION | DATE/TIME |
|---|---|
| | |
| | |
| | |
| | |
| | |
| | |

# BRIMSTONE
*Gonepteryx rhamni*

**FACT FILE**

**SIZE** Wingspan 60mm **HABITAT** Woodland, hedgerows and gardens **LARVAL FOODPLANT** Alder Buckthorn and Buckthorn **FLIGHT PERIOD** Aug–Jun; adults hibernate Sep–Mar

## IDENTIFICATION
Stunning butterfly with a unique wing shape: forewing is hook-tipped, and rounded hindwing has a sharp point near outer edge. Male is brimstone yellow while female is pale whitish yellow; all individuals have a central orange spot on each wing. Underwing and upperwing colours and markings are similar.

## STATUS AND COMMENTS
Common and widespread in central and S Britain, but absent from the N and from much of W Wales. Often visits garden flowers to feed.

**KEY FACT**

A male Brimstone is unmistakable: no other British butterfly is such a uniform, intense yellow. But, in flight, a female recalls a pale Large White; at rest, note its pointed, not rounded, wings and the absence of dark markings.

## SPOTTER'S CHART

| LOCATION | DATE/TIME |
|---|---|
| | |
| | |
| | |
| | |
| | |

# CLOUDED YELLOW

*Colias croceus*

**FACT FILE**

**SIZE** Wingspan 50mm **HABITAT** Open countryside, occasionally gardens **LARVAL FOODPLANT** Lucerne and other pea family members **FLIGHT PERIOD** May–Sep

## IDENTIFICATION

Upperwings are orange-yellow with a dark central spot on forewing; wing margins are dark brown, uniform in male but with yellow spots in female. Underwings are yellow, flushed orange in centre of forewing. Forewing is marked with black spots; hindwing has a row of red dots and a red-ringed central eyespot.

## STATUS AND COMMENTS

An annual immigrant to Britain from S Europe. Numbers vary from year to year: scarce in some years but abundant in others. Cannot survive the British winter.

**KEY FACT** The Clouded Yellow always rests and feeds with its wings closed and underwings exposed. The upperwings are only ever seen in flight: the markings are hard to discern but the intensity of the orange-yellow colour is easily appreciated.

## SPOTTER'S CHART

| LOCATION | DATE/TIME |
| --- | --- |
| | |
| | |
| | |
| | |
| | |
| | |

# SMALL TORTOISESHELL



*Aglais urticae*

**FACT FILE**

**SIZE** Wingspan 40–45mm **HABITAT** Open country, hedgerows and gardens **LARVAL FOODPLANT** Common Nettle **FLIGHT PERIOD** Three broods; flies Mar–Oct and adults hibernate

## IDENTIFICATION

Stunningly colourful butterfly with angular wing margins. Upperwings have striking areas of orange, with yellow and black patches and a blue-spotted dark margin. Underwings are marbled blackish and grey-buff, with a hint of blue spotting on the dark margin; forewing has a broad buffish-yellow patch.

## STATUS AND COMMENTS

Widespread, except in N and upland areas. Once extremely numerous, this species has declined markedly in recent years, although it is still locally common in parts, especially near coasts.

**KEY FACT** If you want to encourage Small Tortoiseshells in the garden, leave patches of Common Nettle for the larvae to eat. Adult butterflies love to feed on Butterfly-bushes and Iceplants.

## SPOTTER'S CHART

| LOCATION | DATE/TIME |
|---|---|
| | |
| | |
| | |
| | |

# PAINTED LADY
*Vanessa cardui*

## FACT FILE

**SIZE** Wingspan 60mm **HABITAT** Open country and sometimes gardens **LARVAL FOODPLANT** Various thistle species **FLIGHT PERIOD** May–Oct

## IDENTIFICATION

Beautiful, fast-flying butterfly with rather pointed forewings. Upperwings are largely pinkish orange with dark spots; forewing tip is dark with white spots. Underside of hindwing is marbled grey-brown and white, and has dark spots; colours and markings on underside of forewing are a paler version of upperside.

## STATUS AND COMMENTS

A migrant species from S Europe that cannot survive the British winter. Numbers vary from year to year but in some seasons the species is very common.

## KEY FACT

The first influx of Painted Lady butterflies usually occurs in May, with more following in midsummer. In some years, the species breeds in Britain, when home-grown individuals boost the number of migrants.

## SPOTTER'S CHART

| LOCATION | DATE/TIME |
|---|---|
| | |
| | |
| | |
| | |
| | |
| | |

# RED ADMIRAL

*Vanessa atalanta*

FACT FILE

**SIZE** Wingspan 60mm **HABITAT** Open countryside, hedgerows and gardens **LARVAL FOODPLANT** Common Nettle **FLIGHT PERIOD** May–Oct

## IDENTIFICATION

Colourful, fast-flying butterfly with angular forewings. Upperwings are mainly black, but forewing has a red central stripe and white patches near tip, while hindwing has a red outer margin. Pattern and colours on underside of forewing are like a faded version of upperwing, while hindwing is marbled blue-black, grey and brown.

## KEY FACT

A Red Admiral larva conceals itself inside a tent formed of Common Nettle leaves bound together with silk. Adult butterflies are fond of feeding on the juices in rotting fruits as well as garden flowers.

## SPOTTER'S CHART

| LOCATION | DATE/TIME |
| --- | --- |
| | |
| | |
| | |
| | |
| | |
| | |

## STATUS AND COMMENTS

Annual migrant from S Europe. Numbers vary from year to year, as does its range, but in most years it is common and widespread except in N Britain.

# PEACOCK
*Inachis io*



**FACT FILE**

SIZE Wingspan 60mm HABITAT Open countryside, hedgerows and gardens LARVAL FOODPLANT Common Nettle FLIGHT PERIOD Jul–Sep, and Mar–May after hibernation

## IDENTIFICATION

Attractive, fast-flying butterfly whose broad wings have jagged margins. Upperwings are maroon overall; forewings have a yellow, maroon and lilac eyespot, and a black patch on leading edge; hindwings have a bold yellow and lilac eyespot. Underwings are marbled smoky grey and brown.

## STATUS AND COMMENTS

Common and widespread except in N and upland Britain; absent from much of N Scotland. A common visitor to gardens, feeding on flowers such as those of the Butterfly-bush.

**KEY FACT**

Adults can be found hibernating in winter in sheds or indoors. They sometimes emerge on sunny days in late winter and are one of the first butterfly species to fly in spring. The larvae feed communally on Common Nettle.

## SPOTTER'S CHART

| LOCATION | DATE/TIME |
|---|---|
| | |
| | |
| | |
| | |

# WHITE ADMIRAL

*Limenitis camilla*

FACT FILE

**SIZE** Wingspan 50mm **HABITAT** Mainly broadleaved woodland, but also nearby mature hedgerows **LARVAL FOODPLANT** Honeysuckle **FLIGHT PERIOD** Jun–Jul

## IDENTIFICATION

Well-marked and distinctive butterfly with a powerful, gliding flight. Forewings are rather pointed while hindwings are rounded. Upperwings are mainly sooty black but marked with a broad, interrupted white band. Underwings are mainly orange-brown but marked with a similar interrupted white band to that seen on upperwings.

## KEY FACT

Usually a wary species, the White Admiral is often easiest to observe when feeding on the flowers of Bramble growing along sunny woodland rides. When drinking nectar, individuals thus preoccupied will sometimes allow a close approach.

## STATUS AND COMMENTS

Locally common moth in suitable habitats in S and central England. Eggs are laid on Honeysuckle that is growing in shady places, and the species occurs only where suitable larval foodplants are growing.

## SPOTTER'S CHART

| LOCATION | DATE/TIME |
|---|---|
| | |
| | |
| | |

# PURPLE EMPEROR
*Apatura iris*



**FACT FILE**

SIZE **Wingspan 65mm** HABITAT **Mature oak woodland** LARVAL FOODPLANT **Goat Willow** FLIGHT PERIOD **Jul–Aug**

## IDENTIFICATION

Large and impressive species with rather pointed forewings and rounded hindwings. Upperwings are marked with white spots; ground colour is brown in female, but male has an amazing bluish-purple sheen in good light. Underwings are brownish with white markings and an eyespot on the forewing.

## STATUS AND COMMENTS

Local and rather rare, being confined to mature oak woodlands in S England. Mature Goat Willows (the larval foodplant) must also be present.

**KEY FACT**

The Purple Emperor is a fast-flying species. Males usually keep to the tops of trees and are a challenge to see well. However, on hot days they sometimes drink from puddles or sup honeydew from leaves.

## SPOTTER'S CHART

| LOCATION | DATE/TIME |
|---|---|
| | |
| | |
| | |
| | |
| | |
| | |

# COMMA

*Polygonia c-album*

**FACT FILE**

**SIZE** Wingspan 45mm **HABITAT** Hedgerows and woods **LARVAL FOODPLANT** Common Nettle and Hop **FLIGHT PERIOD** Two broods: Jun–Jul; Aug–Sep and in Mar–Apr after hibernation

## IDENTIFICATION

Distinctive butterfly with strikingly jagged wing margins. Upperwings are mainly a rich orange-brown; they are marked with dark, squarish spots and have a brown margin. Underwings are marbled grey and brown, and have a conspicuous white 'comma' mark on hindwing.

## STATUS AND COMMENTS

Common and widespread in S and central Britain, mainly in lowland districts; absent from most upland areas and much of the N.

**KEY FACT** Adult butterflies hibernate during the winter months, often among dead leaves or in a crevice in tree bark. In spring and summer, they often visit garden flowers and are also attracted to hedgerow Bramble patches.

## SPOTTER'S CHART

| LOCATION | DATE/TIME |
|---|---|
| | |
| | |
| | |
| | |
| | |
| | |

# MARSH FRITILLARY
*Euphydryas aurinia*

## FACT FILE

**SIZE** Wingspan 40–50mm **HABITAT** Damp, grassy heathland and chalk downland **LARVAL FOODPLANT** Mainly Devil's-bit Scabious **FLIGHT PERIOD** May–Jun

## IDENTIFICATION

Beautiful, slow-flying and rather sluggish butterfly with angled forewings and rounded hindwings. Upperwings are marked with an attractive mosaic pattern of reddish orange, dark brown and buff. Underwings are mainly orange-buff, but with a hint of the upperwing pattern and a pale central band on hindwing.

## KEY FACT

The adult Marsh Fritillary is fond of feeding on knapweed flowers, when it can often be approached closely; it also basks in late-afternoon sunshine with its wings spread flat. The larvae feed gregariously.

## STATUS AND COMMENTS

A distinctly local species whose main centres of distribution are SW England and SW Wales; isolated colonies are found elsewhere in the S, and in W Scotland.

## SPOTTER'S CHART

| LOCATION | DATE/TIME |
|---|---|
| | |
| | |
| | |
| | |

# PEARL-BORDERED FRITILLARY
*Boloria euphrosyne*

**FACT FILE**

**SIZE** Wingspan 40–42mm **HABITAT** Woodland clearings and glades **LARVAL FOODPLANT** Violets **FLIGHT PERIOD** May–Jun

## IDENTIFICATION
Colourful butterfly with rather rounded wings. Upperwings are orange-brown and marked with numerous black spots, some of which align to form rows or jagged lines. Underwings are orange-buff with dark spots on forewing; underwing has seven silver spots on margin and two in centre.

## STATUS AND COMMENTS
A local and declining species, with colonies in S and SW England, and central Scotland; isolated colonies are also found in Wales and N England.

## KEY FACT
The Pearl-bordered Fritillary is a sun-loving butterfly that is very hard to find on dull days. In good weather, it is an active species and often visits Bugle and Primrose flowers to feed.

## SPOTTER'S CHART

| LOCATION | DATE/TIME |
|---|---|
| | |
| | |
| | |
| | |
| | |

# SMALL PEARL-BORDERED FRITILLARY

*Boloria selene*



**FACT FILE**

**SIZE** Wingspan 40mm **HABITAT** Open woodland and scrub, damp grassland and coastal cliffs **LARVAL FOODPLANT** Violets **FLIGHT PERIOD** Jun

## IDENTIFICATION

Attractive butterfly with rather rounded wings. Upperwings are orange-brown and marked with numerous black spots, some of which align to form rows or jagged lines. Underwings are orange-buff with dark spots on forewing; underwing has 7 silver spots on margin and several in the centre.

## STATUS AND COMMENTS

Local and declining butterfly with scattered colonies, mostly in the W half of Britain; SW England, W Wales and W Scotland are the species' strongholds.

**KEY FACT** Very similar to the Pearl-bordered Fritillary. Study the underside of the hind-wing for certain identification: Pearl-bordered has two central silver spots while Small Pearl-bordered has several. There are also subtly different habitat requirements.

## SPOTTER'S CHART

| LOCATION | DATE/TIME |
|---|---|
| | |
| | |
| | |
| | |
| | |
| | |
| | |
| | |

# HIGH BROWN FRITILLARY

*Argynnis adippe*

FACT FILE

**SIZE** Wingspan 60mm **HABITAT** Grassy slopes with scrub, and open woodland **LARVAL FOODPLANT** Violets **FLIGHT PERIOD** Jul–Aug

## IDENTIFICATION

Attractive, fast-flying species. Upperwings are orange-brown and marked with black spots, some of which align to form rows. Underwings are orange-buff with dark spots on forewing and white spots on hindwing. Has a row of white-centred red spots near the hindwing outer margin (not present in Dark Green Fritillary).

## STATUS AND COMMENTS

Local and declining butterfly, now restricted to pockets of suitable habitat in SW England, Wales and NW England; Devon and Cornwall are its strongholds.

KEY FACT

The High Brown Fritillary is an active species that seldom settles for long. Prolonged views of adults can be had while they feed on Bramble and thistle flowers, but despite this preoccupation they are wary and fly off if disturbed.

## SPOTTER'S CHART

| LOCATION | DATE/TIME |
| --- | --- |
| | |
| | |
| | |
| | |

# DARK GREEN FRITILLARY

*Argynnis aglaja*

## FACT FILE

**SIZE** Wingspan 60mm **HABITAT** Flower-rich grasslands, including chalk downland, coastal cliffs and moorland **LARVAL FOODPLANT** Violets **FLIGHT PERIOD** Jul–Aug

## IDENTIFICATION

Attractive, fast-flying butterfly with broad, rounded wings. Upperwings are orange-brown and marked with black spots, some of which align to form rows or jagged lines. Underside of forewing is mostly orange-buff with dark spots; underside of hindwing has greenish-brown scaling and white spots.

## STATUS AND COMMENTS

Widespread but local butterfly that has declined in central and E England. Probably easiest to see on chalk downland, and on moors in W Britain.

## SPOTTER'S CHART

| LOCATION | DATE/TIME |
| --- | --- |
| | |
| | |
| | |
| | |
| | |
| | |

## KEY FACT

Often feeds on thistle and knapweed flowers. Similar to High Brown Fritillary: concentrate on the hindwing underside, which has a greenish-brown (not buffish-brown) ground colour and lacks the row of red spots seen in the High Brown.

# SILVER-WASHED FRITILLARY

*Argynnis paphia*

FACT FILE

**SIZE** Wingspan 60mm **HABITAT** Woodland rides and clearings, and adjacent flowery hedgerows **LARVAL FOODPLANT** Violets **FLIGHT PERIOD** Jun–Aug

## IDENTIFICATION

Fast-flying, broad-winged butterfly with pointed forewings and rounded hindwings. Upperwings are orange-brown, marked with dark streaks and spots on forewing and black spots on hindwing. Underside of forewing is mostly orange-buff, with black spots and a greenish tip; underside of hindwing is greenish with a silvery sheen.

**KEY FACT** Fond of feeding on thistle and Bramble flowers, when it is easy to approach closely. Around 10–15 per cent of Silver-washed Fritillaries have greenish upperwings and pinkish underwings, and are referred to as the form *valesina*.

## STATUS AND COMMENTS

Widespread and locally common in parts of S and central Britain with suitable woodland habitat. Probably easiest to find in S and SW England.

## SPOTTER'S CHART

| LOCATION | DATE/TIME |
| --- | --- |
| | |
| | |
| | |
| | |
| | |
| | |

# HEATH FRITILLARY

*Melitaea athalia*

**FACT FILE**

**SIZE** Wingspan 45mm **HABITAT** Open woodland and grassland with scrub **LARVAL FOODPLANT** Mostly Common Cow-wheat **FLIGHT PERIOD** Jun–Jul

## IDENTIFICATION

Beautifully patterned butterfly with rounded wings. Upperwings are rich orange-brown with a dense network of dark lines. Underside of forewing is orange-buff with a pale tip and variable amounts of dark markings; underside of hindwing has bands of orange-brown and creamy white, demarcated by dark lines.

## STATUS AND COMMENTS

Extremely local and scarce butterfly, with colonies in SE and SW England; those in E Kent and on Exmoor are probably the least vulnerable.

**KEY FACT** The Heath Fritillary flies only on sunny days, and when conditions are overcast it is almost impossible to find. On warm mornings it likes to bask in early sunshine.

## SPOTTER'S CHART

| LOCATION | DATE/TIME |
|---|---|
| | |
| | |
| | |
| | |
| | |
| | |

# GLANVILLE FRITILLARY

*Melitaea cinxia*

## FACT FILE

**SIZE** Wingspan 40mm **HABITAT** Coastal undercliffs and stabilised shingle, and coastal chalk downland **LARVAL FOODPLANT** Sea Plantain **FLIGHT PERIOD** May–Jun

## IDENTIFICATION

Intricately patterned butterfly with slightly angled forewings and rounded hindwings. Upperwings are orange-brown with a network of dark lines. Underside of forewing is orange-buff with a creamy tip and faint dark markings; underside of hindwing has bands of orange and creamy white, demarcated by dark lines.

## STATUS AND COMMENTS

A rare butterfly, confined to the S half of the Isle of Wight, with isolated colonies in S Hampshire. It is invariably found within sight of the sea.

## KEY FACT

The Glanville Fritillary is a sun-loving butterfly that seldom settles for long. However, it will sometimes pause to feed on the flowers of Thrift and will bask with wings spread if a cloud partly obscures the sun.

## SPOTTER'S CHART

| LOCATION | DATE/TIME |
| --- | --- |
| | |
| | |
| | |
| | |

# BLACK HAIRSTREAK

*Satyrium pruni*

## FACT FILE

**SIZE** Wingspan 35mm **HABITAT** Open woodland and adjacent areas of scrub **LARVAL FOODPLANT** Blackthorn **FLIGHT PERIOD** Jul

## IDENTIFICATION

Sluggish butterfly with a tail streamer on each hindwing. Upperwings (seldom seen) are dark brown; both sexes have a row of orange spots on hindwing but female also has spots on forewing. Underwings are rich brown with an orange band, a row of black dots and a white line on both wings.

## KEY FACT

The Black Hairstreak invariably rests and feeds with its wings closed. Typically, it crawls about on the flowers of Wild Privet, or feeds on honeydew on the leaves of its larval foodplant.

## STATUS AND COMMENTS

A local and generally scarce butterfly, restricted to areas of suitable habitat in the East Midlands. Appropriate management of the species' larval foodplant is necessary for its survival.

## SPOTTER'S CHART

| LOCATION | DATE/TIME |
| --- | --- |
| | |
| | |
| | |

# WHITE-LETTER HAIRSTREAK

*Satyrium w-album*

FACT FILE

**SIZE** Wingspan 35mm **HABITAT** Woodland margins and hedgerows where larval foodplant is present **LARVAL FOODPLANT** Elms **FLIGHT PERIOD** Jul–Aug

## IDENTIFICATION

Active butterfly that is difficult to follow in flight. Has a typical hairstreak tail streamer on each hindwing (longer in female than male). Upperwings (seldom seen) are dark brown. Underwings are rich brown with a jagged orange band and white line; on hindwing, the latter forms the letter 'W'.

## STATUS AND COMMENTS

Widespread but local butterfly whose main range is central and S England. Its fortunes declined when Dutch elm disease killed large numbers of elms, its larval foodplant.

## KEY FACT

The White-letter Hairstreak is usually active around treetops, where it is hard to observe well. However, it sometimes visits the flowers of Bramble and thistles, feeding with its wings closed, and allowing observers to get a close look.

## SPOTTER'S CHART

| LOCATION | DATE/TIME |
|---|---|
| | |
| | |
| | |
| | |
| | |
| | |
| | |

# PURPLE HAIRSTREAK
*Neozephyrus quercus*



**FACT FILE**

**SIZE** Wingspan 35–38mm **HABITAT** Oak woodland, and sometimes isolated hedgerow or garden oaks **LARVAL FOODPLANT** Oaks **FLIGHT PERIOD** Jul–Sep

## IDENTIFICATION
Attractive butterfly whose hindwings have a short tail streamer. Upperwings look dark in certain lights; seen well, male has a purple sheen on both wings while in female this colour is restricted to forewings. Underwings are grey with a jagged, dark-bordered white line and orange eyespot, and a small patch on hindwing.

## STATUS AND COMMENTS
Widespread and locally common butterfly in England and Wales, with scattered colonies N to S Scotland. Colonies are entirely dependent on the presence of mature oaks.

## SPOTTER'S CHART

| LOCATION | DATE/TIME |
|---|---|
| | |
| | |
| | |
| | |
| | |
| | |
| | |

**KEY FACT**

The Purple Hairstreak usually flies around the tops of mature oaks, making observation a real challenge. Courtship behaviour, sometimes involving hundreds of individuals, can be seen on sunny evenings and offers a chance to watch the species for extended periods.

# GREEN HAIRSTREAK

*Callophrys rubi*

FACT FILE

**SIZE** Wingspan 25mm **HABITAT** Scrub, grassland, heaths, chalk downs and coasts **LARVAL FOODPLANT** Wide range of plants, but mainly gorse species and Broom **FLIGHT PERIOD** May–Jun

## IDENTIFICATION

Active little butterfly that is hard to follow in flight. Upperwings (seldom seen) are dark brown. Underwings are bright green with faint white marks; always rests with its wings closed, the underwing colour providing perfect camouflage against the foliage on which it rests.

## STATUS AND COMMENTS

Widespread and fairly common, but rather local and also easily overlooked. Found throughout much of England and Wales, with scattered colonies in lowland Scotland.

KEY FACT

Male Green Hairstreaks are territorial and scan for rivals while perched on a prominent spray of vegetation. If you sit patiently, they will often return to the same spot, allowing close and prolonged views.

## SPOTTER'S CHART

| LOCATION | DATE/TIME |
|---|---|
| | |
| | |
| | |
| | |

# BROWN HAIRSTREAK

*Thecla betulae*

**FACT FILE**

SIZE Wingspan 40–50mm HABITAT Blackthorn scrub and hedgerows LARVAL FOODPLANT Blackthorn FLIGHT PERIOD Aug

## IDENTIFICATION

Distinctive butterfly with a short tail streamer on each hindwing. Upperwings are mostly dark brown with orange-tinged tail streamers; female has a broad orange band on forewing. Underwings are orange-brown, marked with fine, dark-edged white lines, and an intense orange flush near tail streamers.

## STATUS AND COMMENTS

A local and generally scarce butterfly with scattered colonies in S England and W Wales. Its decline is linked to the destruction of suitable habitats across its range.

**KEY FACT**

As with other hairstreak species, the larva of the Brown Hairstreak is flattened and rather slug-like. It is bright green and can sometimes be found in spring, feeding on the leaves of its larval foodplant.

## SPOTTER'S CHART

| LOCATION | DATE/TIME |
| --- | --- |
| | |
| | |
| | |
| | |
| | |
| | |

# SMALL COPPER

*Lycaena phlaeas*

FACT FILE

**SIZE** Wingspan 25mm **HABITAT** Flower-rich grassland, heathland and coastal cliffs **LARVAL FOODPLANT** Sheep's Sorrel and Common Sorrel **FLIGHT PERIOD** May–Sep, as 2 or 3 broods

## IDENTIFICATION

Stunningly colourful little butterfly with rather pointed wings. Upperside of forewings are orange with dark spots and a brown margin; upperside of hindwings are brown with an orange marginal band. Underwing pattern recalls that on upperwings, but with muted colours and buff replacing brown.

KEY FACT

The Small Copper is an extremely active butterfly. Occasionally, however, it likes to bask on bare ground or feed for extended periods on flowers such as Wild Thyme, Common Ragwort and Common Fleabane.

## SPOTTER'S CHART

| LOCATION | DATE/TIME |
|---|---|
| | |
| | |
| | |
| | |
| | |
| | |

## STATUS AND COMMENTS

Common and widespread in suitable habitats throughout much of England and Wales. Absent from upland districts in N England, and from N and NW Scotland.

# COMMON BLUE

*Polyommatus icarus*

**FACT FILE**

**SIZE** Wingspan 30–35mm **HABITAT** Flower-rich grassland **LARVAL FOODPLANT** Mainly Common Bird's-foot Trefoil and Black Medick **FLIGHT PERIOD** Two broods: May–Jun and Aug–Sep

## IDENTIFICATION

Attractive little butterfly. Male upperwings are blue with black-edged white margins. Female upperwings are usually brown (sometimes tinged blue in centre) with a row of marginal, dark-centred orange spots. Underwings of both sexes are grey-brown with dark-centred white spots and marginal orange spots on hindwing.

## STATUS AND COMMENTS

Common and widespread in much of lowland Britain, but absent from, or scarce in, mountainous regions and much of N and NW Scotland.

**KEY FACT** This is by far the commonest 'blue' butterfly in the region. It often feeds with its wings open in sunny weather; on dull days, it sometimes rests on plant stems with its wings closed.

## SPOTTER'S CHART

| LOCATION | DATE/TIME |
| --- | --- |
| | |
| | |
| | |
| | |
| | |
| | |
| | |

# CHALKHILL BLUE
*Lysandra coridon*

## FACT FILE

**SIZE** Wingspan 40mm **HABITAT** Chalk and limestone grassland **LARVAL FOODPLANT** Horseshoe Vetch **FLIGHT PERIOD** Jul–Aug

## IDENTIFICATION

Colourful grassland butterfly. Upperwings of male are pale sky blue, with a dark border margin and white fringe. Upperwings of female are brown with a white fringe. Underwings are grey in male, grey-brown in female; in both sexes they are adorned with dark-centred white spots and orange spots.

## SPOTTER'S CHART

| LOCATION | DATE/TIME |
|---|---|
| | |
| | |
| | |
| | |
| | |
| | |

## KEY FACT

The Chalkhill Blue larva is often tended by ants, which protect it from predators in return for a honeydew-like liquid that they relish. The larva feeds mainly at night but can sometimes be found at dusk.

## STATUS AND COMMENTS

Very locally common in S England, where it is entirely restricted to grassland on calcareous soils. Probably easiest to see in the South Downs.

# ADONIS BLUE
*Lysandra bellargus*



**FACT FILE**
SIZE Wingspan 32–34mm HABITAT Chalk downland on S-facing slopes LARVAL FOODPLANT Horseshoe Vetch FLIGHT PERIOD Two broods: May–Jun and Aug–Sep

## IDENTIFICATION
Iconic downland butterfly. Upperwings of male are a dazzling iridescent blue with a chequered white and black fringe. Upperwings of female are brown or brownish blue with a chequered white and black fringe, and with orange spots on hindwing. Underwings of both sexes are grey-buff with dark-centred white spots and orange spots.

## STATUS AND COMMENTS
Very locally common in S England, where it is entirely restricted to calcareous downland with the correct S aspect. Easiest to see in the South Downs.

**KEY FACT**
The Adonis Blue is an active butterfly on sunny days and seldom settles for long. It is easiest to see well in the late afternoon, when it often basks in the last rays of sunshine.

## SPOTTER'S CHART

| LOCATION | DATE/TIME |
|---|---|
| | |
| | |
| | |

# SMALL BLUE
*Cupido minimus*

FACT FILE

**SIZE** Wingspan 25mm **HABITAT** Grassland where the larval foodplant is abundant **LARVAL FOODPLANT** Kidney Vetch **FLIGHT PERIOD** May–Jun

## IDENTIFICATION
An active species, and Britain's smallest 'blue' butterfly. Upperwings are dark grey-brown; male has bluish scaling towards base. Underwings are greyish white, marked with a row of small black dots and flushed greyish blue towards base.

## STATUS AND COMMENTS
Widespread but extremely local and declining butterfly, with small colonies patchily distributed as far N as central coastal Scotland. The species' stronghold is in S England.

### KEY FACT
Small Blues will sometimes drink from the margins of puddles, imbibing salts as well as water in the process. On occasions, they will even drink sweat from the arms of human visitors to their territories.

## SPOTTER'S CHART

| LOCATION | DATE/TIME |
| --- | --- |
| | |
| | |
| | |
| | |

# SILVER-STUDDED BLUE

*Plebeius argus*

## FACT FILE

**SIZE** Wingspan 25–30mm **HABITAT** Mostly heathland, with isolated coastal colonies **LARVAL FOODPLANT** Mainly Heather and gorse species **FLIGHT PERIOD** Jun–Jul

## IDENTIFICATION

Stunning little butterfly. Male has blue upperwings, edged with a chequered black and white fringe. Female's upperwings are brown with a white margin and a marginal row of orange crescents. Underwings of both sexes are grey-brown, with white-ringed black dots and marginal bands of white and orange.

## STATUS AND COMMENTS

Very locally abundant in S England, with the vast majority of colonies found on heathland; here, they are entirely dependent on the correct management of larval foodplants.

## KEY FACT

Away from S heathlands, isolated Silver-studded Blue colonies are also found on limestone grassland on the Isle of Portland and Great Ormes Head, and in dunes on the N Cornish coast.

## SPOTTER'S CHART

| LOCATION | DATE/TIME |
|---|---|
| | |
| | |
| | |
| | |

# BROWN ARGUS

*Aricia agestis*

**FACT FILE**

**SIZE** Wingspan 27–32mm **HABITAT** Grassland, including chalk downs **LARVAL FOODPLANT** Common Rock-rose on chalk **FLIGHT PERIOD** Two broods: Apr–May and Jul–Aug

**KEY FACT** Could be confused with a female Common Blue, but note the Brown Argus's uniformly brown upperwing ground colour (no hint of blue) and striking orange crescents. On the underside of the forewing it has fewer, but larger, dark spots.

## IDENTIFICATION

Attractive little butterfly. Sexes are similar but male is smaller than female. Upperwings are rich brown with a white fringe and submarginal row of orange crescents; has a central black spot on forewing. Underwings are grey-brown with white-ringed black spots and a submarginal band of orange spots.

## STATUS AND COMMENTS

Widespread and very locally common in England and Wales. Some downland colonies have been lost, but the species has spread to new grassland areas where Common Stork's-bill is present.

## SPOTTER'S CHART

| LOCATION | DATE/TIME |
|---|---|
| | |
| | |
| | |
| | |
| | |
| | |

# NORTHERN BROWN ARGUS

*Aricia artaxerxes*



**FACT FILE**

**SIZE** Wingspan 27–32mm **HABITAT** Calcareous grassland **LARVAL FOODPLANT** Common Rock-rose **FLIGHT PERIOD** Jun

## IDENTIFICATION

N counterpart of the Brown Argus. Upperwings are rich brown, with a submarginal row of orange crescents on male's hindwings and spots on both wings of female; Scottish individuals have a central white spot on forewing. Underwings are grey-brown with black-centred white spots and a submarginal band of orange spots.

## STATUS AND COMMENTS

Very locally common in N England and central and E Scotland; its precise distribution is dictated by the presence of its foodplant, which grows only on calcareous soils.

**KEY FACT** The English populations of Northern Brown Argus could be confused with Brown Argus, but the species' ranges and flight times do not overlap. Black markings on the Northern Brown Argus's underwings are always less intense than those on the Brown Argus.

## SPOTTER'S CHART

| LOCATION | DATE/TIME |
|---|---|
| | |
| | |
| | |
| | |
| | |
| | |

# LARGE BLUE
*Maculinea arion*

FACT FILE

**SIZE** Wingspan 30–40mm **HABITAT** Grasslands, on both acid and limestone soils **LARVAL FOODPLANT** Wild Thyme seedheads at first, then ant grubs **FLIGHT PERIOD** Jul

## IDENTIFICATION
An impressive 'blue' butterfly; male is typically smaller than female. Upperwings are pale violet-blue with a blackish outer edge and white fringe. Dark spots are present on forewings; these are largest and darkest in females. Underwings are grey-buff with white-ringed dark spots and a blue flush at base of hindwing.

## STATUS AND COMMENTS
Declared extinct in the UK in 1979, but colonies have since been established in the West Country using butterflies introduced from Sweden. One site – Collard Hill in Somerset, owned by the National Trust – has public access.

## KEY FACT
To survive, the Large Blue needs a habitat where both Wild Thyme and the ant *Myrmica sabuleti* are present. The larva eats ant grubs and in return produces a sugary syrup that is relished by the adult ants.

## SPOTTER'S CHART

| LOCATION | DATE/TIME |
| --- | --- |
| | |
| | |
| | |
| | |
| | |
| | |

# HOLLY BLUE

*Celastrina argiolus*

**FACT FILE** SIZE Wingspan 30mm HABITAT Gardens, hedgerows and woodland edges LARVAL FOODPLANT Holly in spring, Ivy in autumn FLIGHT PERIOD Two broods: Apr–May and Aug–Sep

## IDENTIFICATION

Attractive little butterfly with rather rounded wings. Upperwings are pale violet-blue, with a dark margin that is much broader in females than males, and a chequered white and black fringe in both sexes. Underwings are whitish with small black dots.

## KEY FACT

The Holly Blue needs both Holly and Ivy to complete its annual two-brood life cycle: larvae from eggs laid in spring feed on Holly flowers, while 2nd-brood larvae feed on Ivy flowers.

## STATUS AND COMMENTS

Widespread and common in lowland areas of central and S England, and in lowland parts of Wales. A frequent visitor to gardens where both Holly and Ivy grow.

## SPOTTER'S CHART

| LOCATION | DATE/TIME |
|---|---|
| | |
| | |
| | |

# DUKE OF BURGUNDY

*Hamearis lucina*

FACT FILE

**SIZE** Wingspan 25mm **HABITAT** Calcareous grassland and open woodland **LARVAL FOODPLANT** Mainly Cowslip but also Primrose **FLIGHT PERIOD** May–Jun

## SPOTTER'S CHART

| LOCATION | DATE/TIME |
|---|---|
| | |
| | |
| | |
| | |
| | |
| | |

## IDENTIFICATION

Attractive, well-marked butterfly that resembles a miniature fritillary. Upperwings are patterned with orange and dark brown. Underwings are orange-brown with black markings and with two rows of white spots on hindwing. Overall, the male is usually slightly smaller and darker than female.

## STATUS AND COMMENTS

A rare and declining butterfly whose demise is largely due to habitat destruction. Now restricted to scattered colonies in S England and isolated colonies N to Yorkshire.

### KEY FACT

If you visit a Duke of Burgundy colony before 10am, you stand a chance of seeing the butterflies basking in the early-morning sunshine so that they can warm up; typically, at such times they allow a close approach.

# SPECKLED WOOD

*Pararge aegeria*

## FACT FILE

**SIZE** Wingspan 45mm **HABITAT** Sunny woodland glades and clearings **LARVAL FOODPLANT** Various grasses **FLIGHT PERIOD** Two broods: Apr–Jun and Jul–Sep

## KEY FACT

Male Speckled Woods are territorial and scan their surroundings from sunlit sprays of vegetation; if an intruder is spotted, a spiralling battle ensues in flight, with the potential rival soon being driven off.

## IDENTIFICATION

Charming and alert butterfly with broad, rounded wings. Upperwings are brown, the forewing with pale squarish spots and an eyespot; hindwing has a marginal row of pale spots, some of which contain a dark eyespot. Underwings are grey-brown with a subdued version of the pattern seen on upperwings.

## STATUS AND COMMENTS

A widespread and generally common butterfly in lowland parts of Wales and S and central England; it also occurs in coastal districts of central and W Scotland.

## SPOTTER'S CHART

| LOCATION | DATE/TIME |
|---|---|
| | |
| | |
| | |
| | |
| | |
| | |
| | |

# WALL

*Lasiommata megera*

**FACT FILE**

**SIZE** Wingspan 45mm **HABITAT** Dry grassland, heaths and coasts **LARVAL FOODPLANT** Various grasses, including Wavy Hair-grass **FLIGHT PERIOD** Two broods: May–Jun and Aug–Sep

## IDENTIFICATION

Well-marked, broad-winged butterfly. Upperwings are orange-brown with a network of dark bands; has a single eyespot on forewing and a row of smaller eyespots on hindwing. Underside of forewing is orange-buff with a striking eyespot; underside of hindwing is marbled grey and buff, with indistinct eyespots.

## STATUS AND COMMENTS

Widespread in England and Wales, although the species has declined recently and disappeared from many parts of central and S England. Now only locally common, often near coasts.

**KEY FACT** The Wall loves to bask in the sun with its wings spread on bare ground with dry, free-draining soil, although it seldom settles for long. Its flight is fast and gliding.

## SPOTTER'S CHART

| LOCATION | DATE/TIME |
| --- | --- |
| | |
| | |
| | |
| | |
| | |
| | |

# RINGLET

*Aphantopus hyperantus*

## FACT FILE

**SIZE** Wingspan 45–50mm **HABITAT** Grassland, woodland rides and hedgerows **LARVAL FOODPLANT** Various grasses, including Tufted Hair-grass **FLIGHT PERIOD** Jun–Jul

## IDENTIFICATION

Subtly attractive butterfly with understated colours and markings. Wings are broad and rounded. Upperwings are smoky brown, darker in male than female; has a variable number of eyespots, which are far more conspicuous in female than male. Underwings are yellowish brown with striking white-ringed, dark eyespots.

## STATUS AND COMMENTS

Widespread and locally common in much of England and Wales, and in lowland regions of S and central Scotland. Suffers from agricultural degradation and destruction of its grassland habitats.

## KEY FACT

The Ringlet has a rather slow flight. It likes to feed on thistle flowers, and when preoccupied in this manner it can be approached quite closely. Mating pairs are sometimes found resting on grass stems.

## SPOTTER'S CHART

| LOCATION | DATE/TIME |
|---|---|
| | |
| | |
| | |
| | |
| | |
| | |

# MOUNTAIN RINGLET
*Erebia epiphron*

FACT FILE

**SIZE** Wingspan 32mm **HABITAT** Restricted to mountain grassland **LARVAL FOODPLANT** Mat-grass, an upland grass species **FLIGHT PERIOD** Jun–Jul

## IDENTIFICATION
Small, broad-winged butterfly with slightly pointed forewings and rounded hindwings. Upperwings are dark brown with a submarginal orange band that contains small, dark eyespots. Underwing colours and markings reflect those seen on upperwings, although the colours are muted and paler overall.

## STATUS AND COMMENTS
Entirely restricted to upland mountain slopes in the Lake District and the Scottish Highlands, with colonies found close to where its larval foodplant thrives.

KEY FACT

Your only chances of observing Mountain Ringlets are on sunny days. On dull days, or the second a cloud obscures the sun, the butterflies dive into the cover of dense moorland vegetation and are almost impossible to find.

## SPOTTER'S CHART

| LOCATION | DATE/TIME |
|---|---|
| | |
| | |
| | |

# SCOTCH ARGUS

*Erebia aethiops*

**FACT FILE**

**SIZE** Wingspan 40mm **HABITAT** Lush grassland that includes larval foodplants **LARVAL FOODPLANT** Purple Moor-grass in Scotland, Blue Moor-grass in N England **FLIGHT PERIOD** Jul–Aug

## IDENTIFICATION

Well-marked butterfly with broad, rounded wings. Upperwings are dark brown with an orange submarginal band containing white-centred, dark eyespots; eyespots are variable in size but are typically larger in female than male. Underwings are marbled buff and brown; forewings have an orange band containing eyespots.

## SPOTTER'S CHART

| LOCATION | DATE/TIME |
|---|---|
| | |
| | |
| | |
| | |
| | |

## STATUS AND COMMENTS

Widespread and generally common in suitable habitats in central and W Scotland. Colonies in N England are found on limestone soils.

**KEY FACT**

Although it is most active in sunshine, the Scotch Argus will often continue flying when clouds obscure the sun, and into the early evening in warm weather. It is fond of feeding on Bramble flowers.

# GATEKEEPER

*Pyronia tithonus*

**FACT FILE**

**SIZE** Wingspan 40mm **HABITAT** Hedgerows, and grassland with Bramble patches **LARVAL FOODPLANT** Various grasses, including Timothy and Cock's-foot **FLIGHT PERIOD** Jul–Aug

## IDENTIFICATION

Broad-winged wayside butterfly. Upperwings have patches of orange containing a paired eyespot on forewing, and broad brown margins; extent of orange is greater in female than male. Underside of forewing is similar to upperside, but paler overall; underside of hindwing is mottled grey, brown and buff.

## STATUS AND COMMENTS

Widespread and generally common in lowland districts of Wales and central and S England. The species does best in sheltered, sunny spots.

**KEY FACT** Bramble is the favourite flower of the adult Gatekeeper. It can be seen in hedgerows, but will also visit flowery herbaceous borders in rural gardens. Its flight is rather slow and bouncing.

## SPOTTER'S CHART

| LOCATION | DATE/TIME |
|---|---|
| | |
| | |
| | |
| | |
| | |
| | |
| | |

# MARBLED WHITE

*Melanargia galathea*

## FACT FILE

**SIZE** Wingspan 50mm **HABITAT** Flower-rich grassland **LARVAL FOODPLANT** Various grasses, but particularly Sheep's Fescue **FLIGHT PERIOD** Jul–Aug

## IDENTIFICATION

Almost unmistakable, broad-winged butterfly. Upperwings are white with bold black patches and a network of black lines. Underwing colours and patterns reflect those seen on upperwings, but black markings are greyish; hindwing has eyespots, and white elements of markings are very slightly suffused yellow.

## STATUS AND COMMENTS

Widespread and locally common in S England, with scattered colonies in lowland parts of Wales and central England as far N as Yorkshire.

## SPOTTER'S CHART

| LOCATION | DATE/TIME |
| --- | --- |
| | |
| | |
| | |
| | |
| | |
| | |
| | |

## KEY FACT

The presence of the Marbled White is a good indication that the grassland where it occurs has not been ruined by intensive agricultural practices or so-called 'improvements'. Its flight is rather slow and it spends long periods feeding on thistle flowers.

# GRAYLING

*Hipparchia semele*

FACT FILE

**SIZE** Wingspan 50mm **HABITAT** Dry grassy places, on free-draining soil **LARVAL FOODPLANT** Various grasses, including Sheep's Fescue and Tufted Hair-grass **FLIGHT PERIOD** Jul–Aug

## IDENTIFICATION

Active, broad-winged butterfly that is well camouflaged when resting. Upperwings (never seen at rest) are grey-brown with yellow-buff patches containing dark eyespots. Underside of forewing is yellow-buff with a dark eyespot and marbled grey-brown margin; underside of hindwing is marbled grey, buff and white.

### KEY FACT

The Grayling always rests with its wings closed, often with the hindwing obscuring the forewing. It angles itself towards the sun so that the shadow cast by its wings is minimised, hence making it difficult to spot.

## STATUS AND COMMENTS

Very locally common on heathland and downland in S England. More widespread in coastal districts, from S England to Scotland.

## SPOTTER'S CHART

| LOCATION | DATE/TIME |
| --- | --- |
| | |
| | |
| | |
| | |
| | |
| | |

# MEADOW BROWN

*Maniola jurtina*



**FACT FILE**

**SIZE** Wingspan 50mm **HABITAT** Wide range of grassy habitats **LARVAL FOODPLANT** Wide range of grass species **FLIGHT PERIOD** Jun–Aug

## IDENTIFICATION

Has angular forewings and rounded hindwings. Upperwings are brown overall; male has a small yellow-orange patch and eyespot on forewing, while female has a broad yellow-orange patch and bold eyespot on forewing and subtle yellow patch on hindwing. Underwing patterns and colours reflect those on upperwings, but are much paler overall.

## STATUS AND COMMENTS

Widespread and generally very common butterfly throughout much of lowland England and Wales. Also found in Scotland, but absent from mountainous regions.

**KEY FACT** The Meadow Brown is probably the most familiar and widespread British butterfly. It can be abundant in undisturbed flowery meadows and is a frequent visitor to garden herbaceous borders in rural areas.

## SPOTTER'S CHART

| LOCATION | DATE/TIME |
| --- | --- |
| | |
| | |
| | |
| | |
| | |
| | |
| | |

# SMALL HEATH

*Coenonympha pamphilus*

**FACT FILE**

**SIZE** Wingspan 30mm **HABITAT** Undisturbed grassland **LARVAL FOODPLANT** Wide range of grass species **FLIGHT PERIOD** Two broods: May–Jun and Aug–Sep

## SPOTTER'S CHART

| LOCATION | DATE/TIME |
| --- | --- |
| | |
| | |
| | |
| | |
| | |

## IDENTIFICATION

Has rounded wings. Upperwings (never seen at rest) are orange-buff with a dark border and small eyespot on forewing. Underside of forewing is orange-yellow with a grey tip and margin, and a dark eyespot; underside of hindwing is grey and brown with a creamy-white central band.

## STATUS AND COMMENTS

Widespread grassland species. Locally common, especially in the S, with scattered colonies throughout Wales, central and N England, and lowland Scotland.

**KEY FACT**

The Small Heath always holds its wings closed when feeding or at rest, and its flight is rather slow and fluttering. It has declined markedly in recent years owing to habitat destruction and degradation through agricultural 'improvement'.

# LARGE HEATH

*Coenonympha tullia*



**FACT FILE** SIZE Wingspan 38mm HABITAT Acid upland moors and grassy bogs LARVAL FOODPLANT Mainly Hare's-tail Cottongrass, but also White Beak-sedge FLIGHT PERIOD Jun–Jul

**KEY FACT** The Large Heath always rests with its wings closed. It flies only in sunshine and disappears into dense vegetation on dull days. The species shows considerable geographical variation, each form with its own distinct, recognisable markings.

## IDENTIFICATION

Upland butterfly with broad, rounded wings. Upperwings (never seen at rest) are orange-brown; some geographical forms have eyespots. Underside of forewing is orange-buff with a grey margin and striking eyespot in most forms; underside of hindwing is grey with a pale band and subtle eyespots in most forms.

## STATUS AND COMMENTS

A northern and upland species that is locally common in N Wales, N England, and central, W and N Scotland. Its specific habitat requirements dictate its precise distribution.

## SPOTTER'S CHART

| LOCATION | DATE/TIME |
|---|---|
| | |
| | |
| | |

# CHEQUERED SKIPPER

*Carterocephalus palaemon*

## FACT FILE

**SIZE** Wingspan 25mm **HABITAT** Open birch woodland **LARVAL FOODPLANT** Purple Moor-grass **FLIGHT PERIOD** May–Jun

## IDENTIFICATION

Well-marked little butterfly with angled forewings and rounded hindwings. Upperwings are rich brown with yellow-buff spots arranged in a chequerboard fashion. Underwings are orange-buff with a pattern of dark markings on forewings and pale spots on hindwings.

## STATUS AND COMMENTS

Locally fairly common butterfly in suitable habitats in W Scotland. Formerly occurred in central England but has been extinct there since the 1970s.

## KEY FACT

This little butterfly is active only in sunshine and is fond of basking; it is hard to find on dull days. It often visits flowers such as Bugle to feed, and has a fast, buzzing flight.

## SPOTTER'S CHART

| LOCATION | DATE/TIME |
|---|---|
| | |
| | |
| | |
| | |

# SMALL SKIPPER

*Thymelicus sylvestris*

**FACT FILE**

**SIZE** Wingspan 25mm **HABITAT** Undisturbed grassy meadows **LARVAL FOODPLANT** Various grasses, especially Yorkshire Fog **FLIGHT PERIOD** Jun–Jul

## IDENTIFICATION

Active little butterfly with a rapid, buzzing flight. Upperwings are orange-brown with a dark margin; male has a diagonal dark streak (called a sex brand) on forewing. Underside of forewing is orange-buff with a greenish-grey margin, and hindwing is mostly greenish grey with an orange-buff trailing half.

## STATUS AND COMMENTS

Widespread and common in suitable grassy habitats throughout central and S Britain, becoming scarcer, with more scattered colonies, further N to Northumberland.

**KEY FACT**

The Small Skipper is very similar to the Essex Skipper. To distinguish the two, look at the underside of the tips of the antennae: in the Small they are brown, while in the Essex they are black.

## SPOTTER'S CHART

| LOCATION | DATE/TIME |
|---|---|
| | |
| | |
| | |
| | |

# ESSEX SKIPPER

*Thymelicus lineola*

## FACT FILE

**SIZE** Wingspan 25mm **HABITAT** Undisturbed grassy meadows **LARVAL FOODPLANT** Various grasses, including Creeping Soft-grass and Cock's-foot **FLIGHT PERIOD** Jul–Aug

## KEY FACT

To distinguish this species from the Small Skipper, you need to look at the antennal tips: these are black in the Essex, brown in the Small. Experienced eyes can sometimes determine identity by noting the Essex's overall paler underwings and subtly more heavily marked upperwings.

## SPOTTER'S CHART

| LOCATION | DATE/TIME |
| --- | --- |
| | |
| | |
| | |
| | |
| | |
| | |
| | |

## IDENTIFICATION

Upperwings are orange-brown with a dark margin and slightly darker veins than Small Skipper; male has a diagonal dark streak (called a sex brand) on forewing. Underside of forewing is pale orange-buff with a greenish-grey margin, and hindwing is pale greenish grey with a rather indistinct, pale orange-buff trailing half.

## STATUS AND COMMENTS

Locally common in S England. Its range has expanded in recent years; it now occurs mostly S and E of a line drawn between the Severn and Humber.

# LULWORTH SKIPPER

*Thymelicus acteon*

## FACT FILE

**SIZE** Wingspan 28mm **HABITAT** Coastal grassland **LARVAL FOODPLANT** Tor-grass **FLIGHT PERIOD** Jul–Sep

## IDENTIFICATION

Active skipper with a buzzing flight. Upperwings are olive-brown with a dark margin; forewings have a row of pale spots, arranged like a paw-print, this feature much more striking in female than male. Underwings are pale olive-buff, and with dark patch on trailing edge of hindwing in female.

## KEY FACT

This active species is easiest to see closely in the early morning, before it has become too active to follow. Like other skippers, Lulworth Skipper larvae overwinter in tall grass stems. Cutting for hay, or excessive winter grazing, destroys populations.

## STATUS AND COMMENTS

An extremely local butterfly that is restricted to coastal grassland in Dorset. In suitable locations, however, it can be very common.

## SPOTTER'S CHART

| LOCATION | DATE/TIME |
| --- | --- |
| | |
| | |
| | |
| | |
| | |
| | |

# LARGE SKIPPER

*Ochlodes venatus*

**FACT FILE**

**SIZE** Wingspan 35mm **HABITAT** Undisturbed grassy meadows **LARVAL FOODPLANT** Various grasses, including Purple Moor-grass and Cock's-foot **FLIGHT PERIOD** Jun–Jul

## IDENTIFICATION

Upperwings are orange-brown with a broad, darker margin; male has a diagonal dark streak (called a sex brand) on forewing. Underside of forewing is orange-brown with a greenish margin, pale spots and a dark patch on trailing edge; underside of hindwing is greenish with pale spots and an orange patch on trailing edge.

## SPOTTER'S CHART

| LOCATION | DATE/TIME |
|---|---|
| | |
| | |
| | |
| | |
| | |
| | |

## STATUS AND COMMENTS

Widespread and common across much of lowland England and Wales; it is absent from, or scarce in, upland districts and N England.

**KEY FACT**

The Large Skipper feeds on thistle and knapweed flowers. It also visits rural gardens, attracted by herbaceous borders. Like most other skippers, and unlike the majority of butterflies, it rests with its wings held at an angle.

# SILVER-SPOTTED SKIPPER

*Hesperia comma*

## FACT FILE

**SIZE** Wingspan 35mm **HABITAT** Chalk downland
**LARVAL FOODPLANT** Sheep's Fescue
**FLIGHT PERIOD** Aug–Sep

## IDENTIFICATION

Distinctive, well-marked skipper. Upperwings are orange-brown with broad, darker margins; forewings of male show a diagonal dark stripe and pale spots near tip; forewings of female have more pale spots than male. Underwings are greenish yellow with an orange patch on forewing and silvery spots on hindwing.

## KEY FACT

The Silver-spotted Skipper is an active butterfly that is difficult to follow in flight. Fortunately for observers, it also likes to bask on bare ground and feed on knapweed and thistle flowers.

## STATUS AND COMMENTS

A local butterfly, restricted to chalk downs in S England, mainly on the South Downs and North Downs, but also in the Chilterns.

## SPOTTER'S CHART

| LOCATION | DATE/TIME |
| --- | --- |
| | |
| | |
| | |
| | |

# DINGY SKIPPER

*Erynnis tages*

**FACT FILE**

**SIZE** Wingspan 25mm **HABITAT** Rough meadows and grassy woodland rides **LARVAL FOODPLANT** Mainly Common Bird's-foot Trefoil, but also related vetches **FLIGHT PERIOD** May–Jun

## IDENTIFICATION

Tiny butterfly with muted and understated colours, and a rather moth-like appearance. Flight is buzzing and usually low to the ground. Upperwings are subtly marbled dark brown and smoky grey. Underwings are buffish brown with faint rows of paler buff spots.

## STATUS AND COMMENTS

Widespread in S and central England, with more scattered colonies in Wales and N England, and a few isolated populations in Scotland.

**KEY FACT**

In warm years a 2nd brood of Dingy Skippers is sometimes seen on the wing in Aug. The species likes to bask with its wings spread flat in early-morning sunshine.

## SPOTTER'S CHART

| LOCATION | DATE/TIME |
| --- | --- |
| | |
| | |
| | |
| | |

# GRIZZLED SKIPPER
*Pyrgus malvae*



**FACT FILE**

**SIZE** Wingspan 20mm **HABITAT** Rough grassy habitats and sunny woodland rides **LARVAL FOODPLANT** Mostly Wild Strawberry, cinquefoils and Bramble **FLIGHT PERIOD** May–Jun

## IDENTIFICATION

Well-marked and subtly attractive skipper. Upperwings are dark smoky grey overall, with a chequerboard pattern of squarish white spots and a chequered margin. Underwings are dull greenish yellow with a chequerboard pattern of white spots; these align to form a band on the hindwing.

## STATUS AND COMMENTS

Widespread and very locally common in central and S England; scarce in East Anglia and with a few scattered colonies in Wales.

**KEY FACT** Like the Dingy Skipper, the Grizzled Skipper likes to bask in sunshine with its wings spread flat, often on the leaves of its larval foodplant, Wild Strawberry. However, it rests and roosts with its wings folded.

## SPOTTER'S CHART

| LOCATION | DATE/TIME |
|---|---|
| | |
| | |
| | |
| | |
| | |
| | |

# SMALL MAGPIE

*Eurrhypara hortulata*

**FACT FILE**

**SIZE** Wingspan 25–28mm **HABITAT** Hedgerows, rough grassland and gardens **LARVAL FOODPLANT** Mostly Common Nettle, but sometimes woundworts **FLIGHT PERIOD** Jun–Jul

## IDENTIFICATION

Distinctive moth that rests with its wings spread flat. Upperwings are whitish overall with broad blackish margins and a neat arrangement of black spots. Head, thorax and base of forewings are flushed orange-yellow, and orange-yellow tip to abdomen protrudes beyond wings.

## KEY FACT

The Small Magpie is attracted to light; it is caught in moth traps and rests on walls and fences in the vicinity of outside lights in the garden. It is sometimes disturbed in the daytime from nettle patches.

## SPOTTER'S CHART

| LOCATION | DATE/TIME |
| --- | --- |
| | |
| | |
| | |
| | |
| | |
| | |

## STATUS AND COMMENTS

Widespread and often extremely common moth in S and central Britain, but found mainly in lowland areas where its larval foodplant is common.

# MOTHER OF PEARL

*Pleuroptya ruralis*



**FACT FILE** SIZE Wingspan 28–40mm HABITAT Woodland rides, hedgerows and gardens LARVAL FOODPLANT Common Nettle FLIGHT PERIOD Jun–Jul

## IDENTIFICATION

Rests with its wings spread flat and both forewings and hindwings exposed. In dull light, upperwings look pale buff with reddish-brown scalloped lines and spots. In good light, wings have a pearly, iridescent lustre, with purple, red or green in the sheen depending on the viewing angle.

## STATUS AND COMMENTS

Widespread and common throughout much of lowland Britain, wherever the larval foodplant is common. Probably most numerous in S England.

**KEY FACT** The Mother of Pearl is attracted to light and is often caught in moth traps in good numbers. It can also be seen in the daytime if disturbed into flight from nettle patches.

## SPOTTER'S CHART

| LOCATION | DATE/TIME |
| --- | --- |
| | |
| | |
| | |
| | |

# DECEMBER MOTH

*Poecilocampa populi*

**FACT FILE**

**SIZE** Length 16mm **HABITAT** Deciduous woodland, hedgerows and mature gardens **LARVAL FOODPLANT** Broadleaved trees, including oaks and birches **FLIGHT PERIOD** Nov–Jan

## IDENTIFICATION

Distinctive moth that rests with its wings held in a tent-like manner over body, forewings concealing hindwings. Forewings are mainly sooty black with a pale, jagged cross-line, pale patch at base, and buff and brown chequered margin. Head is pale buff.

**KEY FACT**

The December Moth is attracted to light on mild winter evenings, and can often be found resting on walls in the vicinity of outside lights. Its wing markings and colours are a good match for a dead twig.

## STATUS AND COMMENTS

Widespread and common throughout much of England and Wales, except upland regions; also widespread in lowland Scotland in suitable wooded habitat.

## SPOTTER'S CHART

| LOCATION | DATE/TIME |
|---|---|
| | |
| | |
| | |
| | |

# CHINESE CHARACTER

*Cilix glaucata*



**FACT FILE** SIZE Length 11–13mm HABITAT Hedgerows, woods and gardens LARVAL FOODPLANT Blackthorn, Hawthorn and Wild Crab FLIGHT PERIOD Two broods: May–Jun and Jul–Aug

## IDENTIFICATION

Unmistakable moth that holds its wings in a flattened tent-like manner at rest, with forewings concealing hindwings. Forewings are broad, with a rounded margin and pointed tip; they are whitish overall with a broad orange-brown and bluish-grey central band, and dark marginal scallop markings.

## STATUS AND COMMENTS

Widespread and fairly common across much of England and Wales where the larval foodplants are common; also occurs locally in lowland regions of Scotland.

**KEY FACT** The small Chinese Character often rests conspicuously on the top of a leaf. Its wing markings and posture give it a passing resemblance to a bird dropping, a deception that may cause would-be predators to overlook it.

## SPOTTER'S CHART

| LOCATION | DATE/TIME |
|---|---|
| | |
| | |
| | |
| | |
| | |
| | |

# PEACH BLOSSOM

*Thyatira batis*

FACT FILE

**SIZE** Length 16–18mm **HABITAT** Woodland margins, scrub, hedgerows and gardens **LARVAL FOODPLANT** Bramble **FLIGHT PERIOD** Jun–Jul

## IDENTIFICATION

Unmistakable and attractive moth that sometimes rests with its wings spread flat, although they are usually held in a tent-like manner, forewings concealing hindwings. Forewings are brown with striking pink spots that look like tiny petals. Head and thorax are rather furry.

## KEY FACT

The Peach Blossom is attracted to light, and in the morning it can sometimes be found resting on vegetation in the vicinity of outside lights in gardens where Bramble patches are tolerated or encouraged.

## SPOTTER'S CHART

| LOCATION | DATE/TIME |
|---|---|
| | |
| | |
| | |
| | |
| | |
| | |
| | |

## STATUS AND COMMENTS

Widespread and fairly common moth throughout much of England and Wales; much more local and scarce in Scotland, and usually restricted to lowland and coastal areas.

# BUFF ARCHES

*Habrosyne pyritoides*

**FACT FILE**

**SIZE** Length 19–22mm **HABITAT** Open woodland, hedgerows and gardens **LARVAL FOODPLANT** Bramble and Dewberry **FLIGHT PERIOD** Jun–Aug

## IDENTIFICATION

Distinctive moth that rests with its wings held in a tent-like manner, forewings concealing hindwings. Forewings are divided into two distinct areas, one grey and faintly striated, the other orange-brown and grey, marked with bold white lines and a pattern of swirly pale lines.

**KEY FACT** The Buff Arches' unique markings and patterns are a match for damaged wood: the brown elements are like tree bark, while the striated grey areas are like timber that is freshly exposed when bark is stripped away.

## SPOTTER'S CHART

| LOCATION | DATE/TIME |
| --- | --- |
| | |
| | |
| | |
| | |
| | |

## STATUS AND COMMENTS

Widespread and fairly common in S and central England, and in much of Wales; scarce in N England and rare in, or absent from, much of Scotland, and usually restricted to lowland and coastal areas.

# FIGURE OF EIGHTY

*Tethea ocularis*

FACT FILE

**SIZE** Length 25mm **HABITAT** Woodland, parks and mature gardens **LARVAL FOODPLANT** Aspen and various poplars **FLIGHT PERIOD** May–Jul

## IDENTIFICATION

Appropriately named moth that rests with its wings held in a tent-like manner, forewings concealing hindwings. Forewings are pale brown overall, with a greyish central band defined by black lines and containing a white number '80'. Head and thorax are furry.

## STATUS AND COMMENTS

Widespread and generally common throughout much of England and Wales, occurring N to North Yorkshire and Cumbria. Absent from Scotland.

## KEY FACT

When it is entirely at ease, the Figure of Eighty's wings are rolled slightly in cross section, creating an almost cylindrical appearance. The overall effect is very much like a snapped-off piece of twig.

## SPOTTER'S CHART

| LOCATION | DATE/TIME |
|---|---|
| | |
| | |
| | |
| | |
| | |

# MARCH MOTH

*Alsophila aescularia*

**FACT FILE**

**SIZE** Length 20mm (male), 9mm (female) **HABITAT** Open woodland, parks and gardens **LARVAL FOODPLANT** Broadleaved trees, including oaks and hawthorns. **FLIGHT PERIOD** Mar–Apr

## IDENTIFICATION

Female is wingless and brown. Male is distinctive, with narrow wings that are held flat at rest. Typically, overall outline is triangular, with one forewing overlapping and mostly hiding the other; hindwings are concealed. Forewings are yellow-buff with jagged, dark-edged white cross-lines.

## STATUS AND COMMENTS

Widespread and generally common in well-wooded districts throughout most lowland parts of England, Wales and Scotland.

## SPOTTER'S CHART

| LOCATION | DATE/TIME |
|---|---|
| | |
| | |
| | |
| | |

**KEY FACT**

The male March Moth is the only early-spring species to hold its wings in such a distinctive, overlapping manner. Females can be found by examining the trunks and branches of the larval foodplant trees after dark using a torch.

# LARGE EMERALD

*Geometra papilionaria*

**FACT FILE**

**SIZE** Wingspan 42mm **HABITAT** Woodland, heaths and gardens **LARVAL FOODPLANT** Birches and Hazel **FLIGHT PERIOD** Jul–Aug

## IDENTIFICATION

Stunning, relatively large moth that rests with its wings spread flat in the manner of a butterfly. Forewings have a pointed tip and hindwings are rounded with a wavy margin. Upperwings are bright emerald green with two rows of whitish spots.

## KEY FACT

The wing colour of the Large Emerald is brightest when the moth is newly emerged; it fades with age and wear. The species is attracted to light after dark and very occasionally flies in the daytime.

## STATUS AND COMMENTS

Widespread and fairly common throughout England and Wales, except in upland areas; also occurs in Scotland, in lowland and coastal regions.

## SPOTTER'S CHART

| LOCATION | DATE/TIME |
|---|---|
| | |
| | |
| | |
| | |

# BLOTCHED EMERALD

*Comibaena bajularia*

**FACT FILE**

**SIZE** Wingspan 24–28mm **HABITAT** Mature oak woodland and mature gardens **LARVAL FOODPLANT** Oaks and Hazel **FLIGHT PERIOD** Jun–Jul

## IDENTIFICATION

Colourful and distinctively marked moth. Rests with its wings spread flat in the manner of a butterfly. Upperwings are bright green, with brown and white blotches on trailing edge of forewing and around margins of both wings.

## STATUS AND COMMENTS

Locally fairly common in suitable habitats in S and central England, and in lowland regions of Wales. Absent from N England and Scotland.

**KEY FACT** The markings and colours on the wings of the Blotched Emerald are a good match for a slightly damaged leaf; this camouflage makes it hard to spot when it is resting among foliage. The species is attracted to light.

## SPOTTER'S CHART

| LOCATION | DATE/TIME |
| --- | --- |
| | |
| | |
| | |
| | |
| | |

# BLOOD-VEIN
*Timandra comae*

FACT FILE

**SIZE** Wingspan 31–33mm **HABITAT** Waysides, hedgerows and gardens **LARVAL FOODPLANT** Common Sorrel, docks and Knotgrass **FLIGHT PERIOD** Two broods: May–Jun and Aug–Sep

## IDENTIFICATION
Unmistakable moth that rests with its wings spread flat in the manner of a butterfly. Forewings have a pointed tip and hindwings have an angular margin. Upperwings are yellow-buff, stippled with tiny dark dots; a red transverse line crosses both wings, and both have a red margin.

## KEY FACT
The Blood-vein is an alert and sensitive moth. During the daytime it is easily disturbed from its resting place if you walk through damp wayside vegetation. It is attracted to light after dark.

## STATUS AND COMMENTS
Widespread and fairly common in lowland regions of Wales and S and central England; scarce in, or absent from, much of N England and Scotland.

## SPOTTER'S CHART

| LOCATION | DATE/TIME |
|---|---|
| | |
| | |
| | |
| | |

# RIBAND WAVE

*Idaea aversata*

**FACT FILE**

SIZE **Wingspan 25–30mm** HABITAT **Woodlands, hedgerows and gardens** LARVAL FOODPLANT **Numerous herbaceous plants, including Dandelion and docks** FLIGHT PERIOD **Jun–Aug**

## IDENTIFICATION

Rests with its wings spread flat; forewings are pointed, hindwings are rounded. Occurs as two distinct forms: one has yellow-buff upperwings with a broad brownish band running across both wings; the other has pale buff upperwings with three concentric dark lines running across both wings.

## STATUS AND COMMENTS

Widespread and common across much of England and Wales, except upland areas; more local in Scotland, where it is restricted to lowland districts.

**KEY FACT** The Riband Wave is sometimes found resting on low vegetation in the daytime, and is easily disturbed. It is attracted to light after dark and is often caught in moth traps in good numbers.

## SPOTTER'S CHART

| LOCATION | DATE/TIME |
| --- | --- |
| | |
| | |
| | |
| | |
| | |
| | |

# RED TWIN-SPOT CARPET
*Xanthorhoe spadicearia*

**FACT FILE**

**SIZE** Wingspan 20mm **HABITAT** Woodland, hedgerows, grassland and gardens **LARVAL FOODPLANT** Wide range of low-growing herbaceous plants **FLIGHT PERIOD** May–Aug

## IDENTIFICATION
Attractively marked moth that rests with its wings spread flat, forewings typically largely concealing hindwings. Forewings are reddish buff overall with a darker reddish-brown central band, clearly defined by two paler cross-bands. There are two dark reddish spots near forewing tip; these wear and fade.

**KEY FACT**

The Red Twin-spot Carpet is sometimes disturbed from low-growing vegetation in the daytime, while gardening or walking in the countryside. It is attracted to light and is often caught in moth traps.

## SPOTTER'S CHART

| LOCATION | DATE/TIME |
|---|---|
| | |
| | |
| | |
| | |
| | |

## STATUS AND COMMENTS
Widespread and common across most of England and Wales in a wide range of habitats; scarcer in the N than in the S, and restricted to lowland areas in Scotland.

# BARRED STRAW

*Eulithis pyraliata*



**FACT FILE** SIZE Wingspan 35mm HABITAT Woodland margins, hedgerows, grassland and gardens LARVAL FOODPLANT Common Cleavers and bedstraws FLIGHT PERIOD Jul–Sep

## IDENTIFICATION

Distinctive moth that usually rests with its wings spread flat and wide, but with triangular forewings completely concealing hindwings, and a wide gap between trailing edge of forewing and the abdomen. Forewings are yellow-buff with a broad, darker brown central band.

## STATUS AND COMMENTS

Widespread and common across most of England and Wales except upland regions; also widespread in Scotland, but restricted there to lowland habitats.

## KEY FACT

The Barred Straw can sometimes be disturbed from low-growing vegetation in the daytime; after flying a short distance, it will often rest with its wings folded above its body, in the manner of a roosting butterfly.

## SPOTTER'S CHART

| LOCATION | DATE/TIME |
|---|---|
| | |
| | |
| | |
| | |
| | |
| | |

# COMMON MARBLED CARPET

*Chloroclysta truncata*

FACT FILE

**SIZE** Wingspan 25–30mm **HABITAT** Woodland, hedgerows, parks and gardens **LARVAL FOODPLANT** Shrubs and low-growing plants **FLIGHT PERIOD** Two broods: May–Jun and Aug–Oct

## KEY FACT

This is one of our most variable moths and no one individual will be exactly the same as another. By learning to recognise the two typical forms you will discern characters that help identify mystery individuals.

## STATUS AND COMMENTS

Widespread and generally very common throughout most of England and Wales; also widespread in Scotland, but least numerous in upland regions.

## IDENTIFICATION

Variable moth that rests with its wings spread flat, forewings concealing hindwings. A wide range of forms exists. One typical form has sooty-brown forewings with a broad brown or yellow-brown central band. Another typical form has marbled bands of brown, grey and black with a central whitish band.

## SPOTTER'S CHART

| LOCATION | DATE/TIME |
| --- | --- |
| | |
| | |
| | |
| | |
| | |
| | |
| | |

# RED-GREEN CARPET

*Chloroclysta siterata*

## FACT FILE

**SIZE** Wingspan 30–35mm **HABITAT** Woodland, hedgerows and gardens **LARVAL FOODPLANT** Broadleaved trees and shrubs **FLIGHT PERIOD** Sep–Nov, and Mar–May after hibernation

## IDENTIFICATION

Attractive moth that rests with its wings spread flat, forewings concealing hindwings; the overall outline is triangular. Forewings are variable but usually a rich shade of green, marbled with dark bands, and with a reddish flush along trailing edge and reddish streak towards leading edge.

## STATUS AND COMMENTS

Widespread and locally common in England and Wales wherever suitable wooded habitat occurs; widespread but local in Scotland, favouring wooded lowlands.

## KEY FACT

The precise occurrence of the Red-green Carpet is rather patchy and is undoubtedly linked to the presence or absence of suitable habitat. The species is attracted to light and can be caught in moth traps.

## SPOTTER'S CHART

| LOCATION | DATE/TIME |
|---|---|
| | |
| | |
| | |
| | |
| | |
| | |
| | |
| | |

# JULY HIGHFLYER

*Hydriomena furcata*

**FACT FILE**

**SIZE** Wingspan 25–30mm **HABITAT** Woodland, marshes, grassland, parks and gardens **LARVAL FOODPLANT** Sallows, willows, Hazel and Bilberry **FLIGHT PERIOD** Jul–Aug

## KEY FACT

Although the July Highflyer is variable, the green form is the commonest across most of its range. The green forewing colour is unlike that of other similarly shaped summer moths, making the species readily identifiable.

## IDENTIFICATION

Attractive moth that rests with its wings spread flat, forewings concealing hindwings; the outline is broadly triangular. Forewings variable, but in the typical form they are an unusual and distinctive blue-green with two dark cross-bands. In other forms, green is replaced by yellowish or brown.

## STATUS AND COMMENTS

Widespread and common throughout most of England and Wales, in a wide range of habitats; also widespread in Scotland, even in many upland areas.

## SPOTTER'S CHART

| LOCATION | DATE/TIME |
| --- | --- |
| | |
| | |
| | |
| | |
| | |
| | |

# FOXGLOVE PUG
*Eupithecia pulchellata*

## FACT FILE

**SIZE** Wingspan 18–20mm **HABITAT** Woodland clearings, heaths, cliffs, moors and gardens **LARVAL FOODPLANT** Foxglove **FLIGHT PERIOD** May–Jul

## IDENTIFICATION

Attractive and well-marked moth that rests with its wings spread flat, both forewings and hindwings exposed. Forewings are banded with orange-brown and dark grey. Hindwings are banded with grey and buffish brown. Thorax is orange-brown and abdomen is buffish brown with a dark band.

## STATUS AND COMMENTS

Widespread and locally common in habitats where the larval foodplant thrives, across much of England, Wales and Scotland.

## KEY FACT

The larva of the Foxglove Pug feeds inside the flowers of Foxglove, not on the leaves of the plant; look for it during the summer months. The adult moth is attracted to light and is caught in moth traps.

## SPOTTER'S CHART

| LOCATION | DATE/TIME |
|---|---|
| | |
| | |
| | |

# LIME-SPECK PUG

*Eupithecia centaureata*

## FACT FILE

**SIZE** Wingspan 16–20mm **HABITAT** Meadows, hedgerows, coasts and gardens **LARVAL FOODPLANT** Low-growing plants **FLIGHT PERIOD** May–Jul and Aug–Oct in S. May–Aug in N

## IDENTIFICATION

Distinctive moth that rests with its wings spread flat, hindwings partly hidden by forewings and with gap between wings and body. Forewings are whitish, marbled with buffish brown; inner half of leading edge is dark and has a dark spot midway. Abdomen is dark with a white tip.

## KEY FACT

The patterns and colours of the Lime-speck Pug create the impression of a bird dropping. It is thought that its appearance confers a degree of protection on the moth, the theory being that predators, particularly birds, are likely to avoid it.

## STATUS AND COMMENTS

Widespread and fairly common across much of England and Wales, except upland regions; widespread in Scotland, but restricted there to lowland areas.

## SPOTTER'S CHART

| LOCATION | DATE/TIME |
|---|---|
| | |
| | |
| | |
| | |

# NOVEMBER MOTH

*Epirrita dilutata*



**FACT FILE**

**SIZE** Wingspan 38–42mm **HABITAT** Deciduous woodland, hedgerows, parks and gardens **LARVAL FOODPLANT** Wide range of broadleaved trees and shrubs **FLIGHT PERIOD** Oct–Nov

## IDENTIFICATION

Familiar moth with understated and muted colours and markings. Rests with its wings spread flat, forewings largely covering hindwings. Forewings are variable, but usually overall grey-brown to grey-buff, marked with concentric, jagged dark lines. Hindwings are pale and sometimes project beyond forewings.

## STATUS AND COMMENTS

Widespread and common throughout most of England and Wales, and particularly abundant in wooded habitats; also widespread in Scotland but least numerous in upland districts.

**KEY FACT** This species has two similar cousins (Autumnal and Pale November moths) that are almost impossible to separate without dissection. However, of the three, the November Moth is the most widespread and common, and the mostly likely to be found.

## SPOTTER'S CHART

| LOCATION | DATE/TIME |
|---|---|
| | |
| | |
| | |
| | |
| | |
| | |

# WINTER MOTH
*Operophtera brumata*

FACT FILE

**SIZE** Wingspan 22–26mm **HABITAT** Broadleaved woodland, hedgerows, moors and gardens **LARVAL FOODPLANT** Wide range of deciduous trees and shrubs **FLIGHT PERIOD** Oct–Jan

## SPOTTER'S CHART

| LOCATION | DATE/TIME |
|---|---|
| | |
| | |
| | |
| | |
| | |

## IDENTIFICATION
Winged male rests with its wings spread flat, the rather rounded forewings concealing hindwings. Forewings are pale grey-buff, marked with subtly darker brown bands and dark lines. Female is brown, almost wingless and flightless; she rests on tree bark, concealed in a crevice in the daytime.

## STATUS AND COMMENTS
Widespread and extremely common moth across much of England and Wales, including many upland areas; widespread and common in Scotland, except in mountainous regions.

### KEY FACT
This is the commonest winter-flying moth in most areas and, on mild nights, males are often caught flying in car headlights. To find females, search tree trunks after dark using a torch.

# THE MAGPIE

*Abraxas grossulariata*

## FACT FILE

**SIZE** Wingspan 35–40mm **HABITAT** Hedgerows, moors and gardens **LARVAL FOODPLANT** Shrubs, including hawthorns, currants, and Heather **FLIGHT PERIOD** Jul–Aug

## IDENTIFICATION

Unmistakable moth that rests with its wings spread flat, forewings partly concealing hindwings. Forewings are white with extensive black markings, especially spots that align to form a dark margin, and a dark central band bordering an orange-yellow line. Base of forewing and thorax are marked with orange-yellow.

## SPOTTER'S CHART

| LOCATION | DATE/TIME |
| --- | --- |
| | |
| | |
| | |
| | |
| | |
| | |
| | |

## STATUS AND COMMENTS

Widespread but rather patchily distributed and possibly declining moth in England and Wales; fairly common in lowland Scotland, where Heather is the larval foodplant.

## KEY FACT

The markings and colours of The Magpie are designed to catch the eye and warn potential predators that it is distasteful to eat; even the larva has similar white, black and orange-yellow warning markings.

# SCORCHED WING

*Plagodis dolabraria*

FACT FILE

**SIZE** Wingspan 25–30mm **HABITAT** Woodland, hedgerows, parks and gardens **LARVAL FOODPLANT** Deciduous trees, notably oaks, birches and sallows **FLIGHT PERIOD** May–Jun

## IDENTIFICATION

Well-named moth that rests with its wings spread flat, forewings partly concealing hindwings. Upperwings are marked with fine, concentric dark lines and are flushed dark reddish brown towards outer margin of trailing edges of both wings. Tip of abdomen, which curls upwards, is also flushed reddish brown.

## STATUS AND COMMENTS

Widespread but only locally common in S and central England, and in Wales; rather patchily distributed and scarce in N England and SW Scotland.

## KEY FACT

The Scorched Wing's markings and colours, and its curled, dark-tipped abdomen, give it a passing resemblance to a dry, withered leaf. But, with a bit of imagination, it also resembles a scorched piece of paper.

## SPOTTER'S CHART

| LOCATION | DATE/TIME |
| --- | --- |
| | |
| | |
| | |
| | |
| | |
| | |

# OAK HOOK-TIP
*Watsonalla binaria*

**FACT FILE**

SIZE Wingspan 20–28mm HABITAT Oak woodland, hedgerows, parks and gardens LARVAL FOODPLANT Pedunculate and Sessile oaks FLIGHT PERIOD Two broods: May–Jun and Aug–Sep

## IDENTIFICATION
Distinctive moth that rests with its wings spread flat in the manner of a butterfly. Forewings, with hooked tips, usually partly conceal rounded hindwings. Upperwings of male are orange-brown, female's are yellowish brown; both sexes have two pale cross-lines and two dark central spots on forewing.

## SPOTTER'S CHART

| LOCATION | DATE/TIME |
| --- | --- |
| | |
| | |
| | |
| | |
| | |
| | |

## STATUS AND COMMENTS
Widespread and locally common in central and S England, and in Wales; its occurrence is more patchy in N England and it becomes rare towards the Scottish border.

**KEY FACT**

The Oak Hook-tip is a nocturnal species that is attracted to light. It is sometimes disturbed from overhanging oak branches in the daytime and, in flight, can look rather like a butterfly.

# PEBBLE HOOK-TIP
*Drepana falcataria*

FACT FILE

SIZE Wingspan 28–34mm HABITAT Woodland, wooded heaths, parks and gardens LARVAL FOODPLANT Mainly birches FLIGHT PERIOD Two broods: May–Jun and Aug

## IDENTIFICATION
Attractively marked moth that rests with its wings spread flat, rather like a butterfly. Forewings have hooked tips and sometimes partly conceal rounded hindwings. Forewings are orange-brown with a central eyespot and dark diagonal line running from wingtip. Hindwings are buff and marked with dark lines.

## KEY FACT
In an unusual form of the Pebble Hook-tip, both pairs of wings have a creamy-buff ground colour, and the markings on the hindwing are much bolder and darker than those seen in the commoner orange-brown form.

## STATUS AND COMMENTS
Widespread and fairly common in suitable habitats throughout much of England and Wales; restricted to lowland areas in N England and Scotland.

## SPOTTER'S CHART

| LOCATION | DATE/TIME |
| --- | --- |
| | |
| | |
| | |
| | |

# LILAC BEAUTY

*Apeira syringaria*

## FACT FILE

**SIZE** Wingspan 38–42mm **HABITAT** Broadleaved woodland, parks and gardens **LARVAL FOODPLANT** Honeysuckle, Wild Privet and Ash **FLIGHT PERIOD** Jun–Jul

## IDENTIFICATION

Subtly attractive and distinctive moth. Rests with its wings spread flat but with leading edge of forewings folded. Forewing margin is irregularly angular, while that of hindwing is scalloped. Upperwings are mottled with pale lilac-pink, buff and orange-brown; a dark cross-line bisects both wings.

## STATUS AND COMMENTS

Widespread but only locally common in central and S England, and in Wales; generally rather scarce in N England and S Scotland, and usually found in lowland districts.

## KEY FACT

The folded and creased leading edge to the forewing is unique to the Lilac Beauty. Together with the wing outline and markings, the overall effect is very much like a crumpled leaf, affording the species excellent camouflage.

## SPOTTER'S CHART

| LOCATION | DATE/TIME |
|---|---|
| | |
| | |
| | |
| | |
| | |

# BRIMSTONE MOTH

*Opisthograptis luteolata*

**FACT FILE**

**SIZE** Wingspan 32–36mm **HABITAT** Hedgerows, scrub and gardens **LARVAL FOODPLANT** Mainly Blackthorn and hawthorns **FLIGHT PERIOD** Usually two broods: Apr–Jun and Sep–Oct

## IDENTIFICATION

Colourful and unmistakable moth that rests with its wings spread flat, forewings concealing hindwings. Forewings are bright brimstone yellow with chestnut-brown patches near wingtip and along leading edge; smaller, paler brown markings also adorn the wings.

## STATUS AND COMMENTS

Widespread throughout much of England, Wales and Scotland; sometimes very common in the S of its range where the larval foodplants are abundant.

**KEY FACT**

N populations of the Brimstone Moth may have only one brood a year – in midsummer – while in the far S it can have three broods; here, freshly emerged moths can be found from spring right through to autumn.

## SPOTTER'S CHART

| LOCATION | DATE/TIME |
|---|---|
| | |
| | |
| | |
| | |

# CANARY-SHOULDERED THORN

*Ennomos alniaria*



**FACT FILE**

SIZE **Wingspan 38–40mm** HABITAT **Woodland, hedgerows, parks and gardens** LARVAL FOODPLANT **Deciduous trees, notably birches, limes and Common Alder** FLIGHT PERIOD **Jul–Oct**

## SPOTTER'S CHART

| LOCATION | DATE/TIME |
| --- | --- |
| | |
| | |
| | |
| | |
| | |
| | |

## IDENTIFICATION

**Colourful moth that usually rests with its wings held in a shallow 'V'. Upperwings are orange-brown, with dark stippling; forewings are marked with two dark cross-lines and have a jagged margin. Underwing markings are similar to those on upperwings. Head and thorax are bright yellow and furry.**

## STATUS AND COMMENTS

**Widespread and generally common in suitable habitats throughout much of England and Wales; also widespread in Scotland, where it is commonest in lowland areas.**

**KEY FACT**

**The species' combination of canary-yellow head and 'shoulders' and jagged, leaf-like wings make it almost unmistakable. Seen out of context, it looks striking, but when resting underneath a leaf, it blends in remarkably well.**

# EARLY THORN

*Selenia dentaria*

FACT FILE

**SIZE** Length 20–22mm **HABITAT** Woodland, hedgerows, parks and gardens **LARVAL FOODPLANT** Deciduous shrubs and trees **FLIGHT PERIOD** Two broods: Apr–May and Aug–Sep

## IDENTIFICATION

Well-marked moth that rests with its wings folded over its body; underwings partly conceal forewings. Wing margins are angular and jagged. Underwings are marbled brown and grey, with pale cross-lines; hindwing has orange-brown patches and a pale central 'comma' mark, and forewing has a dark brown tip.

## STATUS AND COMMENTS

Widespread and generally common in a wide range of habitats in England and Wales; also occurs in Scotland, where it is found most commonly in wooded lowland regions.

## SPOTTER'S CHART

| LOCATION | DATE/TIME |
|---|---|
| | |
| | |
| | |
| | |
| | |

KEY FACT

The Early Thorn looks remarkably like a dead leaf when hanging from a twig and resting amongst foliage. It is the only large member of the 'thorn' family of moths to rest with its wings folded above its body.

# PURPLE THORN

*Selenia tetralunaria*

## FACT FILE

**SIZE** Wingspan 30–36mm **HABITAT** Woodland, hedgerows, parks and gardens **LARVAL FOODPLANT** Deciduous trees and shrubs **FLIGHT PERIOD** Two broods: Apr–May and Jul–Aug

## IDENTIFICATION

Rests with its wings held in a shallow 'V'. Forewing margin is jagged, hindwing margin is scalloped. Upperwing and underwing patterns are broadly similar: basal half is reddish brown in 1st brood and orange-buff in 2nd brood, with a broad purplish-grey band outside this. Each wing has a white crescent mark.

## KEY FACT

The Purple Thorn's wing shape and markings give it a leaf-like appearance; when resting suspended from a twig and in dappled shade, the effect is complete. After dark, the species is attracted to light.

## STATUS AND COMMENTS

Widespread and locally common in central and S England, and in Wales; local in N England and scarce in lowland S Scotland.

## SPOTTER'S CHART

| LOCATION | DATE/TIME |
| --- | --- |
| | |
| | |
| | |
| | |
| | |
| | |

# SCALLOPED HAZEL
*Odontopera bidentata*

FACT FILE

**SIZE** Wingspan 32–38mm **HABITAT** Woodland, hedgerows, scrub and gardens **LARVAL FOODPLANT** Wide range of deciduous and coniferous trees and shrubs **FLIGHT PERIOD** May–Jun

## IDENTIFICATION

Rests with its wings spread flat, forewings concealing hindwings. Upperside of forewings ranges in ground colour from buff to greyish buff or brown. A broad band crosses forewing: in pale specimens this appears appreciably darker, while in dark specimens only the band's margins are defined in blackish.

## KEY FACT

The Scalloped Oak has a jagged, scalloped margin to its forewings, which together with the overall colour helps create a resemblance to a dead leaf; this is presumed to serve as camouflage against would-be predators such as birds.

## SPOTTER'S CHART

| LOCATION | DATE/TIME |
| --- | --- |
| | |
| | |
| | |
| | |
| | |

## STATUS AND COMMENTS

Widespread and generally common throughout much of England, Wales and Scotland. The species' eclectic choice of larval foodplants means it is found in a wide range of habitats.

# WAVED UMBER
*Menophra abruptaria*

**FACT FILE**

**SIZE** Wingspan 38–42mm **HABITAT** Woodland, hedgerows, parks and gardens **LARVAL FOODPLANT** Wild and Garden privets, and Common Lilac **FLIGHT PERIOD** Apr–Jun

## IDENTIFICATION

Beautifully marked moth that rests with its wings spread flat, the pointed forewings only partly concealing hindwings. Overall, upperwing ground colour is buff, but numerous fine, dark lines and bands of brown and dark grey create an intricate pattern that resembles freshly cut timber. Very well camouflaged when resting on suitable substrate.

**KEY FACT**

The Waved Umber is good at finding daytime resting places appropriate to its wing markings and colours. New fence panels are sometimes chosen, where the moth's camouflage make it particularly hard to spot.

## STATUS AND COMMENTS

Widespread and fairly common in central and S England, and in Wales; commonest in SE England and scarcest in W Britain.

## SPOTTER'S CHART

| LOCATION | DATE/TIME |
| --- | --- |
| | |
| | |
| | |
| | |
| | |

# SWALLOW-TAILED MOTH

*Ourapteryx sambucaria*

**FACT FILE**

**SIZE** Wingspan 40–50mm **HABITAT** Woodland, hedgerows, parks and gardens **LARVAL FOODPLANT** Ivy, and numerous deciduous trees and shrubs **FLIGHT PERIOD** Jun–Jul

## IDENTIFICATION

Distinctive and unmistakable moth that rests with its wings spread flat, rather like a butterfly. Forewings have pointed, slightly hooked tips and hindwings are sharply angled, with a short tail streamer. Upperwings are pale yellow with two brown cross-lines. Has two brown spots at base of each tail streamer.

## STATUS AND COMMENTS

Widespread and fairly common throughout much of England and Wales, except in upland areas; in Scotland, it is restricted to S lowland regions.

## SPOTTER'S CHART

| LOCATION | DATE/TIME |
|---|---|
| | |
| | |
| | |
| | |
| | |
| | |

**KEY FACT**

The Swallow-tailed Moth is sometimes lured indoors on warm summer evenings, attracted by light through open windows. Its preferred larval foodplant, Ivy, is widespread in gardens, so the moth is common in urban areas.

# BRINDLED BEAUTY

*Lycia hirtaria*



**FACT FILE**

**SIZE** Wingspan 35–40mm **HABITAT** Woodland, hedgerows, parks and gardens **LARVAL FOODPLANT** Deciduous trees, including birches, hawthorns and sallows **FLIGHT PERIOD** Mar–Apr

**KEY FACT**

The male Brindled Beauty is attracted to light and can sometimes be found in the daytime, resting on garden walls and fences near outside lights. Its markings serve as good camouflage when it rests on tree bark.

## IDENTIFICATION

Subtly attractive moth that rests with its wings spread flat, hindwings largely concealed by forewings; outline is rather triangular overall. Forewings are usually grey-buff, but suffused yellow and heavily marked with dark cross-lines and bands. Head and thorax are furry.

## STATUS AND COMMENTS

Widespread and locally common in central and S Britain. Most numerous in SE England and scarcer in the W; patchily distributed and scarce further N.

## SPOTTER'S CHART

| LOCATION | DATE/TIME |
| --- | --- |
| | |
| | |
| | |
| | |
| | |
| | |
| | |

# OAK BEAUTY

*Biston strataria*

FACT FILE

**SIZE** Wingspan 30–40mm **HABITAT** Woodland, hedgerows, parks and gardens where oaks are present **LARVAL FOODPLANT** Deciduous trees and shrubs **FLIGHT PERIOD** Mar–Apr

## IDENTIFICATION

Attractive and distinctively marked moth that rests with its wings spread flat, hindwings often largely concealed by forewings. Forewings are marked with bands of white (speckled with black) and reddish brown. Hindwings are similarly marked but usually hidden. Male's antennae are feathery.

## KEY FACT

The colours and markings of the Oak Beauty allow it to blend in well when resting on tree bark. The species is attracted to light and is often caught in moth traps in areas where it is common.

## STATUS AND COMMENTS

Widespread and locally common in S Britain. Most numerous in SE England; locally and generally scarce further N and W.

## SPOTTER'S CHART

| LOCATION | DATE/TIME |
|---|---|
| | |
| | |
| | |
| | |
| | |
| | |

# PEPPERED MOTH

*Biston betularia*

**FACT FILE**

**SIZE** Wingspan 40–55mm **HABITAT** Woodland, hedgerows, parks and gardens **LARVAL FOODPLANT** Deciduous trees and shrubs, and low-growing plants **FLIGHT PERIOD** May–Aug

**KEY FACT**

The Peppered Moth rests in sites that suit the camouflage afforded by its wing markings: speckled white forms rest on lichen-covered tree bark; dark forms rest underneath shaded branches, often where lichens are absent.

## IDENTIFICATION

Rests with its wings spread flat, forewings often partly concealing hindwings. Forewings are narrow and pointed, and overall outline is often broadly triangular. Usually occurs in two distinct forms: one has white upperwings that are strongly speckled and stippled with black; the other is uniformly sooty black. Intermediate forms sometimes also occur.

## STATUS AND COMMENTS

Widespread and generally common throughout much of England and Wales; also occurs locally in Scotland, where it is restricted to lowland areas, mainly in the S.

## SPOTTER'S CHART

| LOCATION | DATE/TIME |
| --- | --- |
| | |
| | |
| | |
| | |

# WILLOW BEAUTY

*Peribatodes rhomboidaria*

FACT FILE

**SIZE** Wingspan 30–40mm **HABITAT** Woodland, hedgerows, parks and gardens **LARVAL FOODPLANT** Wide range of deciduous trees and shrubs **FLIGHT PERIOD** Jul–Aug

## IDENTIFICATION

Intricately marked moth that rests with its wings spread flat, rather like a butterfly; forewings are triangular, hindwings are rounded. Upperwings are usually greyish brown, but are strongly stippled with black and have dark cross-lines, the central one of which is broadest and darkest.

## KEY FACT

The Willow Beauty's upperwings are a perfect match for certain types of tree bark. The wing margins are often pressed so flat to the substrate on which it is sitting that little or no telltale shadow can be seen.

## SPOTTER'S CHART

| LOCATION | DATE/TIME |
| --- | --- |
| | |
| | |
| | |
| | |
| | |
| | |

## STATUS AND COMMENTS

Widespread and generally common across much of Wales and central and S England; occurs in N England and S Scotland, where it is restricted to lowlands.

# LIME HAWK-MOTH

*Mimas tiliae*

**FACT FILE**

**SIZE** Wingspan 65mm **HABITAT** Hedgerows, parks and gardens **LARVAL FOODPLANT** Limes, but also birches, Common Alder and elms **FLIGHT PERIOD** May–Jun

## IDENTIFICATION

Unmistakable moth that rests with its wings spread flat; forewings are narrow with a jagged margin and usually conceal hindwings. Upperside of forewings is pinkish buff with two deep green central spots and a paler olive-green margin. Head and sides of thorax are dark green and abdomen is pale green.

## STATUS AND COMMENTS

Widespread and locally common in S and central England and Wales; scarce and very local in N England, and absent from Scotland.

**KEY FACT**

The Lime Hawk-moth is caught in moth traps in small numbers. Its larva is pale green with a diagonal pale stripe and red dot on each segment, and a 'horn' at the tail end.

## SPOTTER'S CHART

| LOCATION | DATE/TIME |
| --- | --- |
| | |
| | |
| | |
| | |
| | |
| | |
| | |
| | |

# EYED HAWK-MOTH

*Smerinthus ocellata*

FACT FILE

**SIZE** Wingspan 75–80mm **HABITAT** Hedgerows, parks and gardens **LARVAL FOODPLANT** Willows, sallows and apples **FLIGHT PERIOD** May–Jul

## IDENTIFICATION

Unmistakable moth that rests with its wings spread flat. If moth is undisturbed, its marbled pinkish-buff and dark brown forewings conceal hindwings. If disturbed, forewings are spread and body arches, revealing blue eyespot with red surround on hindwings. Centre of thorax is blackish, flanked by pale grey.

## STATUS AND COMMENTS

Widespread and fairly common in Wales and central and S England, except in upland areas; scarce and local in N England and absent from Scotland.

## KEY FACT

The eye markings on the Eyed Hawk-moth's hindwings are 'flashed' suddenly when the moth is disturbed, which is thought to act as a deterrent against birds, and perhaps mammalian predators too.

## SPOTTER'S CHART

| LOCATION | DATE/TIME |
| --- | --- |
| | |
| | |
| | |
| | |
| | |

# POPLAR HAWK-MOTH
*Laothoe populi*

## FACT FILE

**SIZE** Wingspan 65–75mm **HABITAT** Scrub, hedgerows, parks and gardens **LARVAL FOODPLANT** Poplars, Aspen and sallows **FLIGHT PERIOD** May–Jul

## IDENTIFICATION

Rests with wings spread flat. Forewings are mostly pinkish grey with a broad brown central band and brown on margin. If moth is undisturbed, hindwings partly project in front of forewings and show pinkish grey and brown. If it is disturbed, forewings are spread and body arches, revealing red spot on hindwings.

## STATUS AND COMMENTS

Widespread and common throughout much of Wales and S and central England; widespread and locally common in N England and Scotland, but mostly in lowland areas.

## SPOTTER'S CHART

| LOCATION | DATE/TIME |
| --- | --- |
| | |
| | |
| | |
| | |
| | |
| | |

## KEY FACT

When resting peacefully, hanging from a branch or twig, the Poplar Hawk-moth looks like a dead leaf. When it is disturbed, the red spot revealed on the hindwings is thought to act as a deterrent to would-be predators.

# PRIVET HAWK-MOTH

*Sphinx ligustri*

FACT FILE

**SIZE** Wingspan 95–100mm **HABITAT** Hedgerows, scrub, parks and gardens **LARVAL FOODPLANT** Privets, Common Lilac and Ash **FLIGHT PERIOD** Jun–Jul

## IDENTIFICATION

Impressively large and unmistakable moth that rests with its wings spread flat. Narrow, pointed forewings are buffish grey with dark lines and marbling; they usually partly conceal hindwings, which are pink with dark bands. Thorax is dark grey-brown and abdomen has pink and blackish bands on side.

## KEY FACT

The species is particularly common on calcareous soils. Look for its impressive larvae (bright green with diagonal purple stripes and a black 'horn') on stunted Wild Privet bushes in coastal dunes and downland scrub in Aug.

## SPOTTER'S CHART

| LOCATION | DATE/TIME |
|---|---|
| | |
| | |
| | |
| | |
| | |

## STATUS AND COMMENTS

Widespread and fairly common in Wales and central and S England, and most numerous in lowland areas and around coasts; absent further N.

# HUMMINGBIRD HAWK-MOTH

*Macroglossum stellatarum*

**FACT FILE** SIZE Wingspan 45–50mm HABITAT Coastal habitats, parks and gardens LARVAL FOODPLANT Various bedstraw species FLIGHT PERIOD May–Oct

## IDENTIFICATION

Invariably seen in flight, when its brown forewings and orange hindwings are seen as a blur. White patches on the sides of the otherwise mainly blackish abdomen are easily noted in hovering individuals. Rests on ground or walls with forewings concealing hindwings; well camouflaged and hard to spot.

**KEY FACT**

This day-flying species is easy to recognise but hard to follow as it is constantly on the move. It uses its long proboscis to drink nectar from the flowers of Red Valerian, Butterfly-bush and other garden plants.

## STATUS AND COMMENTS

Annual immigrant from S Europe, seen from spring to autumn but with main influx in Jul and Aug. Commonest on S coast and in S England.

## SPOTTER'S CHART

| LOCATION | DATE/TIME |
| --- | --- |
| | |
| | |
| | |
| | |
| | |

# ELEPHANT HAWK-MOTH
*Dielephila elpenor*

**FACT FILE**

SIZE Wingspan 65–70mm HABITAT Hedgerows, rough grassland, woodland clearings and gardens LARVAL FOODPLANT Willowherbs FLIGHT PERIOD May–Jun

## IDENTIFICATION

Stunningly attractive moth that rests with its wings spread flat, forewings partly concealing hindwings. Forewings are pink and yellow-olive, with two cross-lines defining and enclosing a pale pink band. Hindwings are pinkish, and head, thorax and abdomen are marked with forewing colours.

**KEY FACT**

The Elephant Hawk-moth is named after its larva, which is usually brown (sometimes green), and with a head end that fancifully resembles an elephant's trunk; the eyespots are thought to deter would-be predators.

## SPOTTER'S CHART

| LOCATION | DATE/TIME |
|---|---|
| | |
| | |
| | |
| | |
| | |

## STATUS AND COMMENTS

Widespread and generally common in central and S England, and in Wales; scarce in N England; in Scotland, restricted to lowlands and coastal districts, mainly in the S.

# BUFF-TIP

*Phalera bucephala*

**FACT FILE**

SIZE Length 30mm HABITAT Woodland, hedgerows, parks and gardens LARVAL FOODPLANT Various deciduous trees, including birches, limes and oaks FLIGHT PERIOD May–Jul

## IDENTIFICATION

Well-marked and unmistakable moth that rests with its wings folded in a tight tent-like manner; forewings conceal hindwings. Forewings are silvery grey with a dark cross-line near base, and a near-circular buffish-yellow patch near tip. Thorax is silvery grey and front of head is buff.

## STATUS AND COMMENTS

Widespread and generally common moth throughout much of England and Wales; less numerous in Scotland, where it is mostly restricted to lowland and W regions.

## KEY FACT

The unusual markings and colours of the Buff-tip give it an uncanny resemblance to a snapped piece of twig, and one of Silver Birch in particular. This camouflage presumably helps the moth avoid detection by predators.

## SPOTTER'S CHART

| LOCATION | DATE/TIME |
|---|---|
| | |
| | |
| | |
| | |
| | |

# PUSS MOTH

*Cerura vinula*

## FACT FILE

**SIZE** Length 35mm **HABITAT** Open woodland, scrub, parks and gardens **LARVAL FOODPLANT** Willows, sallows, Aspen and poplars **FLIGHT PERIOD** May–Jul

## IDENTIFICATION

Striking and unmistakable moth that rests with its wings held in a tent-like manner, forewings concealing hindwings. Forewings are white, intricately marked with fine black lines, and with orange veins. Head, thorax and legs are white, marked with black spots, and extremely furry.

## KEY FACT

The Puss Moth larva is an amazing creature. It is squat and green, with a broad head, a dark dorsal band, and two whip-like appendages at the tail end that it uses to ward off parasitic flies.

## STATUS AND COMMENTS

Widespread and generally common moth throughout most of England and Wales; also occurs in Scotland, but mostly absent from upland areas.

## SPOTTER'S CHART

| LOCATION | DATE/TIME |
| --- | --- |
| | |
| | |
| | |
| | |
| | |
| | |

# SALLOW KITTEN
*Furcula furcula*



**FACT FILE**

**SIZE** Length 20mm **HABITAT** Open woodland, scrub and hedgerows **LARVAL FOODPLANT** Sallows and willows **FLIGHT PERIOD** Two broods: May–Jun and Jul–Aug

## IDENTIFICATION

Rests with its wings held in a tent-like manner, forewings concealing hindwings. Forewings are whitish overall, stippled with black dots and with a broad, orange-tinged central band that is bordered by black and orange. There is a row of marginal black dots. Head, thorax and legs are furry.

## STATUS AND COMMENTS

Widespread and fairly common in central and S Britain, but commonest in the SE; local in Scotland, where it is restricted mainly to lowland areas.

**KEY FACT**

The Sallow Kitten larva is squat and angular, with a pale dorsal band and two whip-like tail appendages. It can be found feeding on the leaves of sallows, often on surprisingly small, spindly shrubs.

## SPOTTER'S CHART

| LOCATION | DATE/TIME |
| --- | --- |
| | |
| | |
| | |
| | |
| | |
| | |

# LOBSTER MOTH

*Stauropus fagi*

FACT FILE

SIZE Length 32mm HABITAT Woodland, hedgerows, parks and gardens LARVAL FOODPLANT Mainly oaks, Beech and Hazel FLIGHT PERIOD May–Jul

## IDENTIFICATION

Rests with its wings held in a shallow tent-like manner, forewings concealing hindwings. Forewings are reddish grey, marked with a jagged, pale cross-line and a row of pale black crescent-shaped spots near margin. Base of trailing edge is flushed red. Head, thorax and legs are furry.

## STATUS AND COMMENTS

Widespread and locally common in Wales and S and central England, mostly in the vicinity of mature deciduous woodland; absent from N England and Scotland.

KEY FACT

The larva of the Lobster Moth is truly bizarre and gives the species its name. When mature, it has a curled-over abdomen tip and long legs, giving it a fanciful resemblance to a lobster.

## SPOTTER'S CHART

| LOCATION | DATE/TIME |
| --- | --- |
| | |
| | |
| | |
| | |
| | |

# PEBBLE PROMINENT

*Eligmodonta ziczac*



**FACT FILE**

**SIZE** Length 25mm **HABITAT** Open woodland, parks and gardens **LARVAL FOODPLANT** Aspen, poplars, sallows and willows **FLIGHT PERIOD** May–Jun

## IDENTIFICATION

Distinctive moth that rests with its wings held in a tent-like manner, forewings concealing hindwings. Forewings are buffish brown overall, but with a bold fingernail-like patch near tip marked with black, brown and white; trailing edge of wing has a white patch.

### KEY FACT

The fingernail-like marking on the forewing of the Pebble Prominent looks a bit like a knot on the branch or twig of a tree; this affords the moth good camouflage when it rests in a suitable spot.

## SPOTTER'S CHART

| LOCATION | DATE/TIME |
|---|---|
| | |
| | |
| | |
| | |
| | |

## STATUS AND COMMENTS

Widespread and generally common throughout most of England and Wales; also widespread in Scotland, but commonest there in lowland regions and around coasts, mainly in the S and W.

# IRON PROMINENT
*Notodonta dromedarius*

FACT FILE

**SIZE** Length 25–28mm **HABITAT** Woodland, hedgerows and gardens **LARVAL FOODPLANT** Mainly birches and Common Alder **FLIGHT PERIOD** Two broods: May–Jun and Jul–Aug

## IDENTIFICATION
Colourful moth that rests with its wings held in a tent-like manner, forewings concealing hindwings; wing 'prominences' are seen in profile. Forewings are sooty grey overall, but richly marked with rusty-brown and yellow bands and lines. Head, thorax and front legs are furry.

## KEY FACT
Although the colours and markings on the forewings of the Iron Prominent look striking when the moth is seen in isolation, its camouflage is extremely good when it rests on an appropriately coloured tree branch.

## SPOTTER'S CHART

| LOCATION | DATE/TIME |
|---|---|
| | |
| | |
| | |
| | |
| | |

## STATUS AND COMMENTS
Widespread and generally common moth throughout most of England and Wales; also widespread in Scotland, but commonest there in lowland regions.

# SWALLOW PROMINENT
*Pheosia tremula*

## FACT FILE

**SIZE** Length 28–30mm **HABITAT** Woods, parks and gardens **LARVAL FOODPLANT** Willows, sallows and poplars **FLIGHT PERIOD** Apr–May and Jul–Aug in S. Jul–Aug in N

## IDENTIFICATION

Well-marked moth that rests with its wings held in a tent-like manner, forewings concealing hindwings. Forewings are buff overall, palest towards leading edge and grading to brown towards trailing edge, which is dark. Dark coloration extends and fades around margin, emphasising several narrow white wedge marks.

## STATUS AND COMMENTS

Widespread and generally common throughout most of England and Wales; rather local in Scotland, where it is commonest in lowland regions.

## KEY FACT

To separate this species from the superficially similar Lesser Swallow Prominent, study the white wedge markings on the forewing margin: in Swallow Prominent there are two or three of these, while in Lesser Swallow Prominent there is just one.

## SPOTTER'S CHART

| LOCATION | DATE/TIME |
|---|---|
| | |
| | |
| | |
| | |
| | |

# LESSER SWALLOW PROMINENT

*Pheosia gnoma*

**FACT FILE**

**SIZE** Length 26–28mm **HABITAT** Woodland, wooded heathland, parks and gardens **LARVAL FOODPLANT** Silver and Downy birches **FLIGHT PERIOD** Two broods: Apr–May and Jun–Aug

## IDENTIFICATION

Well-marked moth that rests with its wings held in a tent-like manner, forewings concealing hindwings. Forewings are buff overall, palest towards leading edge and grading to brown towards trailing edge, which is dark. Dark coloration extends and fades around margin, emphasising a single white wedge mark.

**KEY FACT**

To separate this species from Swallow Prominent, look at the white wedge markings on the forewing (see that species for details). The Lesser Swallow Prominent also tends to have 'warmer' brown forewings than its cousin.

## STATUS AND COMMENTS

Widespread and generally common throughout most of England and Wales; also widespread in Scotland, where it is commonest in lowland regions.

## SPOTTER'S CHART

| LOCATION | DATE/TIME |
| --- | --- |
| | |
| | |
| | |
| | |
| | |
| | |

# COXCOMB PROMINENT

*Ptilodon capucina*

**FACT FILE**

**SIZE** Length 25mm **HABITAT** Woodland, hedgerows, parks and gardens **LARVAL FOODPLANT** Deciduous trees including birches, Hazel and oaks **FLIGHT PERIOD** May–Jun and Aug

## IDENTIFICATION

Distinctive moth that rests with its wings held in a tight tent-like manner, forewings concealing hindwings. Forewings are rather uniformly rich brown with darker veins, and have a curved and scalloped outer margin. Seen in profile, the 'prominences' are obvious. Dorsal surface of thorax has striking white hairs.

## STATUS AND COMMENTS

Widespread and generally common throughout much of England and Wales; also widespread in Scotland, where it is most numerous in lowland areas.

**KEY FACT**

When resting, the Coxcomb Prominent looks just like a dead leaf – sitting on the woodland floor, it is extremely hard to spot. If you pick the moth up it will 'play dead', reinforcing its inanimate appearance.

## SPOTTER'S CHART

| LOCATION | DATE/TIME |
|---|---|
| | |
| | |
| | |
| | |
| | |

# PALE PROMINENT

*Pterostoma palpina*

## FACT FILE

**SIZE** Length 30mm **HABITAT** Woodland, hedgerows, parks and gardens **LARVAL FOODPLANT** Aspen, poplars and willows **FLIGHT PERIOD** Usually two broods: May–Jun and Jul–Aug

## IDENTIFICATION

Rests with its wings held in a tight tent-like manner, forewings concealing hindwings. At rest, palps project at head end, tufted abdomen tip projects beyond wings at tail end, and shows a row of prominences on back. Wings are greyish buff with dark lines and veins.

## KEY FACT

When resting, the Pale Prominent looks for all the world like a piece of woodchip. This deception, designed to fool would-be predators, is made all the more real as the moth 'plays dead' when touched.

## STATUS AND COMMENTS

Widespread and common in S and central England, and in Wales; more local in N England; restricted mainly to lowlands and coastal areas in Scotland.

## SPOTTER'S CHART

| LOCATION | DATE/TIME |
|---|---|
| | |
| | |
| | |
| | |
| | |
| | |

# CHOCOLATE-TIP
*Clostera curtula*

**FACT FILE** SIZE Length 17mm HABITAT Woodland, hedgerows, parks and gardens LARVAL FOODPLANT Aspen, poplars, sallows and willows FLIGHT PERIOD Usually two broods: Apr–May and Aug–Sep

## IDENTIFICATION
Beautiful and unmistakable moth that rests with its wings held in a tight tent-like manner, forewings concealing hindwings. Forewings are silvery grey to grey-buff, with white cross-lines and a reddish-brown patch at tip. Front of head and tip of abdomen are reddish brown.

## STATUS AND COMMENTS
Widespread and locally common in S-central and S England; in Wales, its range is restricted to the S; elsewhere, the species is scarce or absent.

## KEY FACT
Although its markings are distinctive, the Chocolate-tip is surprisingly well camouflaged when resting on a twig. The abdomen is usually concealed, but if the moth is agitated it is sometimes raised above the wings.

## SPOTTER'S CHART

| LOCATION | DATE/TIME |
| --- | --- |
| | |
| | |
| | |
| | |
| | |
| | |
| | |

# FIGURE OF EIGHT

*Diloba caeruleocephala*

FACT FILE

**SIZE** Length 20–22mm **HABITAT** Woodland, hedgerows, parks and gardens **LARVAL FOODPLANT** Blackthorn, hawthorns, apples and related shrubs **FLIGHT PERIOD** Oct–Nov

## IDENTIFICATION

Well-marked moth that rests with its wings held in a tent-like manner, forewings concealing hindwings. Forewings are brown overall, with two dark central cross-lines containing a white marking that often resembles the number '8'. Head and thorax are furry, and antennae of male are feathered.

## STATUS AND COMMENTS

Widespread and generally common in Wales and central and S England, least so in the W; local in N England; restricted to lowland areas in Scotland.

KEY FACT

The Figure of Eight is the only autumn-flying moth you are likely to come across with such distinctive forewing markings. It is attracted to light and is often caught in moth traps.

## SPOTTER'S CHART

| LOCATION | DATE/TIME |
| --- | --- |
| | |
| | |
| | |
| | |
| | |
| | |

# THE VAPOURER
*Orgyia antiqua*

**FACT FILE** SIZE Length 16mm (male), 8mm (female) HABITAT Woodland, hedgerows, heaths, parks and gardens LARVAL FOODPLANT Deciduous trees and shrubs FLIGHT PERIOD Jul–Sep

## IDENTIFICATION
Male rests with wings spread flat, forewings concealing hindwings. Forewings are orange-brown, marked with faint dark lines, and have a dark-framed white spot towards outer margin. Male's antennae are feathery (see photograph). Female is wingless and brown, with a plump, swollen abdomen.

**KEY FACT** The wingless female Vapourer attracts a mate by emitting a chemical called a pheromone. She usually mates shortly after emerging from the pupa, and her batch of eggs is often laid on the pupal cocoon.

## STATUS AND COMMENTS
Widespread and common throughout much of England and Wales, although commonest in the S; also widespread in Scotland, but absent from uplands.

## SPOTTER'S CHART

| LOCATION | DATE/TIME |
|---|---|
| | |
| | |
| | |
| | |
| | |
| | |

# YELLOW-TAIL

*Euproctis similis*

**FACT FILE**

**SIZE** Length 23–25mm **HABITAT** Woodland, hedgerows, scrub and gardens **LARVAL FOODPLANT** Mainly hawthorns and Blackthorn **FLIGHT PERIOD** Jul–Aug

## IDENTIFICATION

Rests with its wings folded flat against one another, or in an acutely angled tent-like manner; forewings obscure hindwings. Forewings are pure white; head, thorax and most of abdomen are also pure white and hairy. Tip of abdomen is yellow, and is raised and exposed if moth is agitated.

**KEY FACT** The Yellow-tail should come with a health warning: hairs from both the adult moth and the larva can cause skin irritation if touched, or breathing difficulties if inhaled. To avoid this, do not handle the moth or its larva.

## STATUS AND COMMENTS

Widespread and common in much of Wales and central and S England, except in upland areas; local in N England and S Scotland.

## SPOTTER'S CHART

| LOCATION | DATE/TIME |
|---|---|
| | |
| | |
| | |
| | |

# WHITE ERMINE

*Spilosoma lubricipeda*

**FACT FILE**

**SIZE** Length 28mm **HABITAT** Woods, hedgerows, parks and gardens **LARVAL FOODPLANT** Herbaceous plants, including docks and Common Nettle **FLIGHT PERIOD** May–Jul

## IDENTIFICATION

Attractive and distinctive moth that rests with its wings held in a tent-like manner, forewings concealing hindwings. Forewings are white with variable numbers of black dots. Head and thorax are extremely furry. Yellow and black abdomen is occasionally revealed if moth is disturbed.

## STATUS AND COMMENTS

Widespread and common throughout most parts of England, Wales and Scotland; found in a wide range of sites, including many upland habitats.

**KEY FACT**

The White Ermine is a distinctive moth that is sometimes found resting on shady wayside vegetation during the daytime. It is attracted to light after dark and is often caught in moth traps.

## SPOTTER'S CHART

| LOCATION | DATE/TIME |
| --- | --- |
| | |
| | |
| | |

# ROSY FOOTMAN

*Miltochrista miniata*

FACT FILE

SIZE **Length 14mm** HABITAT **Open woodland, hedgerows and gardens** LARVAL FOODPLANT **Lichens, especially Dog Lichen** FLIGHT PERIOD **Jul–Aug**

## IDENTIFICATION

Attractive and almost unmistakable moth that rests with its wings spread flat, forewings concealing hindwings. Forewings have a curved outer margin and pointed tip; they are rosy pink with black markings, including a basal cross-line, an extremely wavy central cross-line, and a row of dots near margin.

## SPOTTER'S CHART

| LOCATION | DATE/TIME |
| --- | --- |
| | |
| | |
| | |
| | |
| | |
| | |

## STATUS AND COMMENTS

Widespread and locally common in S England and S Wales; local and rather scarce in central England; absent elsewhere.

## KEY FACT

The preferred foodplants of the Rosy Footman's larva are lichens that grow on tree trunks and branches. Consequently, the species is usually most numerous in mature woodlands and coastal districts where lichens thrive.

# COMMON FOOTMAN
*Eilema lurideola*



**FACT FILE**

**SIZE** Length 25mm **HABITAT** Woodland, hedgerows, parks and gardens **LARVAL FOODPLANT** Lichens growing on trees **FLIGHT PERIOD** Jul–Aug

**KEY FACT**

The Scarce Footman is a similar species that, despite its name, is locally common. Unlike the Common Footman, which holds its wings rather flat, the wings are usually held in a slightly rolled manner; and the yellow stripe on the forewing leading edge does not taper.

## IDENTIFICATION

Unusual moth that rests with its wings folded flat over body, one of the long, narrow forewings concealing the other, and both hindwings. Forewings are mostly grey, with a narrow yellow stripe along leading edge that tapers towards tip. Head and legs are also yellow.

## STATUS AND COMMENTS

Widespread and common in S and central England and Wales; local and scarcer in N England; restricted mainly to coasts in Scotland.

## SPOTTER'S CHART

| LOCATION | DATE/TIME |
|---|---|
| | |
| | |
| | |
| | |
| | |
| | |

# GARDEN TIGER

*Arctia caja*

FACT FILE

**SIZE** Wingspan 65mm **HABITAT** Open grassy habitats, scrub, coasts and gardens **LARVAL FOODPLANT** Wide range of low-growing herbaceous plants **FLIGHT PERIOD** Jul–Aug

## IDENTIFICATION

Impressive and colourful moth that rests with its wings held flat; typically, forewings conceal hindwings but when the moth is alarmed, forewings are spread and hindwings revealed. Forewings are white with variable brown patches; hindwings are orange with variable blue spots. Head is red and abdomen is orange.

## KEY FACT

The colours on the hindwings are revealed when the Garden Tiger is disturbed and serve to startle would-be predators. The larva is extremely hairy and consequently often referred to as a 'woolly bear'.

## STATUS AND COMMENTS

Widespread and still very locally common throughout, but has declined markedly in recent years; now absent from many suburban areas where it was once common.

## SPOTTER'S CHART

| LOCATION | DATE/TIME |
|---|---|
| | |
| | |
| | |
| | |
| | |
| | |

# RUBY TIGER

*Phragmatobia fuliginosa*

**FACT FILE**

**SIZE** Length 22mm **HABITAT** Grassland, woodland, parks and gardens **LARVAL FOODPLANT** Low-growing plants **FLIGHT PERIOD** May–Jun and Aug–Sep in S. May–Jul in N

## IDENTIFICATION

Subtly colourful moth that rests with its wings spread flat, forewings usually concealing hindwings. Forewings are a dull ruby red with a dark central dot. Hindwings are seen in flight and sometimes revealed in an alarmed resting moth; they are bright red with a dark marginal band.

**KEY FACT**

The Ruby Tiger larva is extremely hairy and is often seen walking, at surprising speed, across roads and tracks in warm weather in summer. Adult moths from Scotland are a duller red than S individuals.

## SPOTTER'S CHART

| LOCATION | DATE/TIME |
|---|---|
| | |
| | |
| | |
| | |
| | |
| | |
| | |

## STATUS AND COMMENTS

Widespread and generally common throughout much of England and Wales; also occurs in Scotland, where it is widespread in upland moorland habitats as well as lowlands.

# PALE TUSSOCK

*Calliteara pudibunda*

FACT FILE

**SIZE** Length 30mm **HABITAT** Woodland, hedgerows, parks and gardens **LARVAL FOODPLANT** Wide range of deciduous trees and shrubs **FLIGHT PERIOD** May–Jun

## IDENTIFICATION

Attractive moth that usually rests with its wings held in a shallow tent-like manner, forewings concealing hindwings. Forewings are pale grey to greyish buff, with a subtly darker central band and slightly darker grey scallops near margin. Head, thorax and front legs are extremely hairy.

**KEY FACT** The larva of the Pale Tussock is easily recognised, being covered in long yellow tufts of hairs on each segment, with black in between; there is also a long tuft of reddish hairs at the tail end.

## STATUS AND COMMENTS

Widespread and common throughout much of Wales and central and S England; more local and rather scarce in N England; absent from Scotland.

## SPOTTER'S CHART

| LOCATION | DATE/TIME |
|---|---|
| | |
| | |
| | |
| | |
| | |

# BLACK ARCHES

*Lymantria monacha*

## FACT FILE

**SIZE** Length 22–24mm **HABITAT** Broadleaved woodland, hedgerows and gardens **LARVAL FOODPLANT** Mainly oaks, but sometimes other deciduous trees **FLIGHT PERIOD** Jul–Aug

## IDENTIFICATION

Well-marked moth that rests with its wings spread flat, forewings concealing hindwings; outline is broadly triangular. Forewings are white, but are strongly marked with several jagged, wavy black lines; margins are marked with a series of black dots. Pinkish abdomen is sometimes revealed if moth is agitated.

## STATUS AND COMMENTS

Widespread but distinctly local in Wales and S and S-central England; scarce further N in central England; absent from N England and Scotland.

## SPOTTER'S CHART

| LOCATION | DATE/TIME |
| --- | --- |
| | |
| | |
| | |
| | |
| | |

## KEY FACT

Black Arches is a striking species that can sometimes be found resting on tree trunks and wooden fences in the daytime. It is attracted to light after dark and is often caught in moth traps.

# THE CINNABAR

*Tyria jacobaeae*

FACT FILE

**SIZE** Length 21–23mm **HABITAT** Open grassy habitats, roadside verges and gardens **LARVAL FOODPLANT** Ragworts **FLIGHT PERIOD** May–Jul

## IDENTIFICATION

Unmistakable moth that usually rests with its wings held in a shallow tent-like manner, forewings concealing hindwings. Forewings are sooty grey with red stripes on leading and trailing edges, and two red spots on margins. Hindwings are red and sometimes revealed if moth is agitated.

## STATUS AND COMMENTS

Widespread and generally common throughout Wales and S and central England; rather scarce in N England; found mainly in lowland areas in Scotland.

### KEY FACT

The larva of The Cinnabar is poisonous, and its orange and black stripes warn would-be predators of this. The adult moth is sometimes disturbed into flight in the daytime and is attracted to light after dark.

### SPOTTER'S CHART

| LOCATION | DATE/TIME |
|---|---|
| | |
| | |
| | |
| | |
| | |
| | |

# LARGE YELLOW UNDERWING

*Noctua pronuba*

**FACT FILE**

**SIZE** Length 25mm **HABITAT** Wide range of open habitats, including gardens **LARVAL FOODPLANT** Wide range of low-growing herbaceous plants **FLIGHT PERIOD** Jun–Sep

## IDENTIFICATION

Rests with its wings held flat, one forewing overlapping the other and hiding hindwings. Forewings are marbled brown, with a dark kidney-shaped spot, pale rounded spot and dark wedge near tip of leading edge; inner half of leading edge is usually pale. Hindwings are orange-yellow with a rather narrow black band.

**KEY FACT**

The Large Yellow Underwing is one of our commonest moths. It is alert and is easily disturbed into flight during the daytime. It is attracted to light after dark and is often caught in moth traps.

## SPOTTER'S CHART

| LOCATION | DATE/TIME |
| --- | --- |
| | |
| | |
| | |
| | |
| | |

## STATUS AND COMMENTS

Widespread and often extremely common moth throughout most of England, Wales and Scotland, least so in upland areas. Migrants from mainland Europe boost resident numbers.

# BROAD-BORDERED YELLOW UNDERWING

*Noctua fimbriata*

FACT FILE

SIZE Length 25mm HABITAT Open woodland, hedgerows, grassland and gardens LARVAL FOODPLANT Wide range of low-growing herbaceous plants FLIGHT PERIOD Jul–Sep

## IDENTIFICATION

Well-marked moth that rests with its wings held flat, one forewing overlapping the other and hiding hindwings. Forewings are marked with pale and dark bands; buffish orange overall in female, rich brown in male. Hindwings are orange-yellow with a broad black submarginal band.

## KEY FACT

The Broad-bordered Yellow Underwing often rests in low vegetation in the daytime and is alert and easily disturbed; as it scurries away or flies for a short distance, the broad black border on its hindwings is obvious.

## SPOTTER'S CHART

| LOCATION | DATE/TIME |
| --- | --- |
| | |
| | |
| | |
| | |
| | |

## STATUS AND COMMENTS

Widespread and often very common throughout much of England and Wales, least so in upland areas; widespread in Scotland, but restricted mainly to lowlands.

# LESSER YELLOW UNDERWING

*Noctua comes*



**FACT FILE**

**SIZE** Length 20–23mm **HABITAT** Open woodland, hedgerows, grassy places and gardens **LARVAL FOODPLANT** Wide range of herbaceous plants **FLIGHT PERIOD** Jul–Sep

## IDENTIFICATION

Rests with its wings held flat, one forewing partly overlapping the other and hiding hindwings. Forewings vary from grey to brown overall, typically with dark kidney-shaped and oval spots and dark cross-lines. Hindwing is orange-yellow with a narrow submarginal dark band and a dark central crescent.

## STATUS AND COMMENTS

Widespread and often abundant in lowland England and Wales; also widespread and locally common throughout much of Scotland, including upland areas.

**KEY FACT**

The Lesser Yellow Underwing is appreciably smaller than other common underwing species. Like its relatives, however, the colours on the hindwings are thought to startle would-be predators when they are revealed suddenly.

## SPOTTER'S CHART

| LOCATION | DATE/TIME |
| --- | --- |
| | |
| | |
| | |
| | |

# HEART AND CLUB

*Agrotis clavis*

FACT FILE

**SIZE** Length 19–22mm **HABITAT** Open grassy habitats, waste ground, coastal dunes and gardens **LARVAL FOODPLANT** Wide range of herbaceous plants **FLIGHT PERIOD** Jun–Jul

## IDENTIFICATION

Rests with its wings held flat, one forewing mostly overlapping the other and hiding hindwings. Forewing ground colour varies from buffish to brown, with a dark heart-shaped spot, a blunt club-shaped mark near wing base, and a smaller dark spot. Head and thorax are same general colour as forewing.

## KEY FACT

The Heart and Club is usually commonest in gardens and near coasts in S England. It is attracted to light after dark and is often caught in moth traps. Its precise occurrence is rather local.

## SPOTTER'S CHART

| LOCATION | DATE/TIME |
| --- | --- |
| | |
| | |
| | |
| | |
| | |

## STATUS AND COMMENTS

Widespread and generally common in S and S-central England, and in Wales; rather local in N England and Scotland, where it is found mostly in lowlands.

# HEART AND DART

*Agrotis exclamationis*

**FACT FILE** SIZE Length 20–22mm HABITAT Open woodland, grassy habitats, hedgerows and gardens LARVAL FOODPLANT Wide range of low-growing herbaceous plants FLIGHT PERIOD May–Jul

## IDENTIFICATION

Well-marked moth that rests with its wings held flat, one forewing partly overlapping the other and hiding hindwings. Forewing is usually grey-brown with a bold black heart-shaped marking and a tapering black dart-like marking near base. A paler oval mark and faint scalloped cross-lines are also present.

## STATUS AND COMMENTS

Widespread and extremely common in much of Wales and central and S England; less numerous in N England; in Scotland, commonest in lowlands.

**KEY FACT** The Heart and Dart has a black half-collar on the front of the thorax, seen most obviously when viewed head-on; in the superficially similar Heart and Club, the thorax is uniform brown.

## SPOTTER'S CHART

| LOCATION | DATE/TIME |
| --- | --- |
| | |
| | |
| | |
| | |
| | |
| | |

# THE FLAME
*Axylia putris*

**FACT FILE**

SIZE Length 16–18mm HABITAT Wide range of open lowland habitats, including gardens LARVAL FOODPLANT Wide range of herbaceous plants FLIGHT PERIOD Jun–Jul

**KEY FACT**

The markings and almost cylindrical appearance of a resting Flame give it a resemblance to a snapped-off piece of twig or a dead stem. It is attracted to light and is often caught in moth traps.

## IDENTIFICATION
Distinctive and well-marked moth that rests with its wings rolled tightly around body, forewings concealing hindwings. Forewings are buffish brown overall, marked with dark dots, a dark spot, and dark veins and cross-lines. Thorax is dark brown and head is buff.

## STATUS AND COMMENTS
Widespread and generally common in S and central England, and in Wales; local in N England; found mainly in lowlands and near coasts in Scotland.

## SPOTTER'S CHART

| LOCATION | DATE/TIME |
|---|---|
| | |
| | |
| | |
| | |
| | |
| | |
| | |

# SETACEOUS HEBREW CHARACTER

*Xestia c-nigrum*



## FACT FILE

SIZE Length 17–22mm HABITAT Grassland, gardens and allotments LARVAL FOODPLANT Wide range of herbaceous plants FLIGHT PERIOD May–Jun in S. Jul–Aug in N

## IDENTIFICATION

Well-marked moth that rests with its wings held flat, one forewing partly overlapping the other and hiding hindwings. Forewings are brown overall, but with a rather triangular, pale buff mark on the leading edge that is sharply bordered by a black saddle-shaped marking. There are also faint cross-lines.

## KEY FACT

On the face of it, this species appears very similar to the Hebrew Character. However, that species is seen mostly in Mar and early Apr, so the flying times of the two moths do not overlap.

## STATUS AND COMMENTS

Widespread and common, or abundant, in much of England and Wales, least so in upland districts; commonest in lowland areas in N England and Scotland.

## SPOTTER'S CHART

| LOCATION | DATE/TIME |
|---|---|
| | |
| | |
| | |
| | |

# SIX-STRIPED RUSTIC

*Xestia sexstrigata*

FACT FILE

**SIZE** Length 18–20mm **HABITAT** Damp grassy habitats, hedgerows and gardens **LARVAL FOODPLANT** Wide range of low-growing herbaceous plants **FLIGHT PERIOD** Jul–Aug

## IDENTIFICATION

Well-marked moth that rests with its wings held flat, one forewing partly overlapping the other and concealing hindwings. Forewings are buffish brown, marked with fine dark veins, fine dark cross-lines, and a dark-framed pale oval spot. Head appears rather pointed when viewed from above.

## KEY FACT

The fine dark lines that mark its forewings are the Six-striped Rustic's best identification features, and there are few species with which it can be confused. It is often caught in moth traps.

## STATUS AND COMMENTS

Widespread and locally common throughout much of Wales and S and central England; more local in N England; favours lowlands in Scotland.

## SPOTTER'S CHART

| LOCATION | DATE/TIME |
|---|---|
| | |
| | |
| | |
| | |
| | |
| | |

# THE GOTHIC

*Naenia typica*

**FACT FILE** **SIZE** Length 20–22mm **HABITAT** Damp meadows and woodland rides, shady hedgerows and gardens **LARVAL FOODPLANT** Wide range of herbaceous plants **FLIGHT PERIOD** Jun–Jul

## IDENTIFICATION

Distinctive moth that rests with its wings held flat, forewings concealing hindwings. Outline is broadly triangular. Forewings are grey-brown and strikingly marked with a network of white veins and cross-lines; has bold, white-framed oval and kidney-shaped spots, between which is a dark patch.

## SPOTTER'S CHART

| LOCATION | DATE/TIME |
| --- | --- |
| | |
| | |
| | |
| | |
| | |
| | |
| | |

## KEY FACT

Seldom comes to light and only occasionally caught in moth traps. Sometimes seen feeding at heavily scented garden flowers after dark, and occasionally disturbed from low-growing damp vegetation in the daytime.

## STATUS AND COMMENTS

Widespread but only locally common in England and Wales, mainly in lowland areas; local in Scotland, where it is confined to lowland and coastal regions.

# GREY ARCHES

*Polia nebulosa*

**FACT FILE**

**SIZE** Length 23–28mm **HABITAT** Broadleaved woodland, mature hedgerows and rural gardens **LARVAL FOODPLANT** Deciduous trees and shrubs **FLIGHT PERIOD** Jun–Jul

## IDENTIFICATION

Subtly attractive moth that rests with its wings spread flat, forewings concealing hindwings. Outline is broadly triangular. Forewings are grey, with a powdery appearance created by tiny, dark stippling. There are dark-framed kidney-shaped, rounded and oval markings, and dark marbling and mottling.

## KEY FACT

Grey Arches is attracted to light and is caught in moth traps in small numbers. It can sometimes be found in the daytime resting on shaded fence panels in the garden.

## STATUS AND COMMENTS

Widespread and generally common in central and S England; more local in N England; restricted mainly to lowland and coastal areas in Scotland.

## SPOTTER'S CHART

| LOCATION | DATE/TIME |
| --- | --- |
| | |
| | |
| | |
| | |
| | |

# BRIGHT-LINE BROWN-EYE
*Lacanobia oleracea*



**FACT FILE** SIZE Length 17–20mm HABITAT Hedgerows, scrub, meadows and gardens LARVAL FOODPLANT Wide range of low-growing herbaceous plants FLIGHT PERIOD May–Jul

## IDENTIFICATION

Descriptively named moth that rests with its wings spread flat, forewings concealing hindwings. Outline is broadly triangular. Forewings are a rich brown with a jagged white cross-line near the margin, a white-framed oval marking, and a white-framed orange-brown kidney-shaped marking.

## STATUS AND COMMENTS

Widespread and often extremely common in much of S and central England, and in Wales; widespread and common in N England and Scotland, least so in upland regions.

**KEY FACT** The Bright-line Brown-eye is often most numerous in gardens, where its larvae feed on a range of cultivated plants. It is attracted to light after dark, and is often caught in moth traps.

## SPOTTER'S CHART

| LOCATION | DATE/TIME |
| --- | --- |
| | |
| | |
| | |
| | |
| | |
| | |
| | |

# MARBLED CORONET

*Hadena confusa*

**FACT FILE**

**SIZE** Length 16–18mm **HABITAT** Calcareous grassland and coasts **LARVAL FOODPLANT** Seeds of Bladder Campion, Sea Campion and related plants **FLIGHT PERIOD** May–Jun

## IDENTIFICATION

Rests with its wings spread flat, forewings concealing hindwings. Outline is triangular, and overall appearance can be almost black and white. Forewing ground colour is blackish, stippled with fine white dots. Has white patches at wingtip, halfway along leading edge, at base of wings, and on head and thorax.

## KEY FACT

The wonderful and striking markings of the Marbled Coronet make it stand out against a plain background. However, when resting among lichens – on a rock, for example – the moth blends in beautifully.

## STATUS AND COMMENTS

Widespread and locally common in S England in both coastal and inland areas; mainly coastal elsewhere in England, Wales and Scotland.

## SPOTTER'S CHART

| LOCATION | DATE/TIME |
|---|---|
| | |
| | |
| | |
| | |
| | |
| | |

# THE SPRAWLER

*Asteroscopus sphinx*



## FACT FILE

**SIZE** Length 20–24mm **HABITAT** Woodland, hedgerows and gardens **LARVAL FOODPLANT** Deciduous trees and shrubs **FLIGHT PERIOD** Oct–Dec

## IDENTIFICATION

Well-marked moth that rests with its wings spread flat or in a shallow tent-like manner, forewings concealing hindwings. Outline is narrowly triangular. Forewings are buffish brown with orange-brown suffusion and numerous dark streaks. The thorax is marked with two dark central streaks that converge at front.

## STATUS AND COMMENTS

Widespread and locally common in suitable habitats in S and central England and Wales; local and patchily distributed in N England; absent from Scotland.

## KEY FACT

There are no other similar-looking moths flying so late in the year with which The Sprawler can be confused. It is attracted to light after dark and is often caught in moth traps.

## SPOTTER'S CHART

| LOCATION | DATE/TIME |
| --- | --- |
| | |
| | |
| | |
| | |
| | |

# THE LYCHNIS
*Hadena bicruris*

## FACT FILE

**SIZE** Length 16–19mm **HABITAT** Hedgerows, woodland margins and rural gardens **LARVAL FOODPLANT** Seeds of Red Campion and related plants **FLIGHT PERIOD** Jun–Jul

## IDENTIFICATION

Rests with its wings spread flat or in a shallow tent-like manner, forewings concealing hindwings. Forewings are brown overall with white cross-lines, the central ones converging and enclosing a pale 'V' marking that embraces white-framed oval and kidney-shaped markings, and inside which is a dark triangle that abuts leading edge.

## KEY FACT

In the daytime, The Lychnis hides away at the base of low-growing vegetation, where its wing markings blend in surprisingly well with the tangle of dry stems and leaves. It is sometimes caught in moth traps.

## STATUS AND COMMENTS

Widespread and generally common throughout much of England, Wales and Scotland in habitats where its larval foodplants thrive; absent from many upland areas.

## SPOTTER'S CHART

| LOCATION | DATE/TIME |
|---|---|
| | |
| | |
| | |
| | |
| | |
| | |

# FEATHERED GOTHIC

*Tholera decimalis*

## FACT FILE

**SIZE** Length 19–24mm **HABITAT** Grassland and woodland rides **LARVAL FOODPLANT** Various grasses, including Sheep's Fescue **FLIGHT PERIOD** Aug–Sep

## IDENTIFICATION

Well-marked moth that rests with its wings spread flat or in a shallow tent-like manner, forewings concealing hindwings. Outline is triangular. Forewings are brown to grey-brown overall, with bold white veins, and white-ringed rounded and kidney-shaped markings near middle of leading edge.

## STATUS AND COMMENTS

Widespread and generally common in central and S England, and in Wales; more local in N England and Scotland, where it is restricted mainly to lowlands.

## KEY FACT

The Feathered Gothic is superficially similar to The Gothic, but it flies later in the year and the leading edge of its forewing is straight rather than curved. The male Feathered Gothic has extremely feathered antennae.

## SPOTTER'S CHART

| LOCATION | DATE/TIME |
|---|---|
| | |
| | |
| | |
| | |
| | |
| | |

# SMALL QUAKER
*Orthosia cruda*

FACT FILE

**SIZE** Length 14–17mm **HABITAT** Woodland, hedgerows and gardens **LARVAL FOODPLANT** Range of deciduous trees and shrubs **FLIGHT PERIOD** Mar–Apr

## IDENTIFICATION
Rests with its wings spread in shallow tent-like manner, forewings concealing hindwings. Forewing ground colour varies from grey-buff stippled with black dots, to reddish brown; has a dark kidney-shaped spot halfway along, near leading edge. A faint, pale-ringed rounded marking, and a pale cross-line parallel to outer margin, are sometimes also present.

## STATUS AND COMMENTS
Widespread and generally common in central and S England, and in Wales; more local in N England and Scotland, where it is restricted to wooded lowlands.

## SPOTTER'S CHART

| LOCATION | DATE/TIME |
|---|---|
| | |
| | |
| | |
| | |
| | |
| | |

**KEY FACT**

Together with the Common Quaker, this is one of the commonest moths of those that emerge in early-spring. Large numbers are often caught in moth traps at night, and the species can sometimes be disturbed from wayside vegetation during the daytime.

# COMMON QUAKER
*Orthosia cerasi*



**FACT FILE**

**SIZE** Length 15–19mm **HABITAT** Woodland, hedgerows and gardens **LARVAL FOODPLANT** Deciduous trees and shrubs **FLIGHT PERIOD** Mar–Apr

## IDENTIFICATION
Rests with its wings spread flat or in a shallow tent-like manner, forewings concealing hindwings. Forewing ground colour varies from buff to reddish brown. Some individuals show a dark cross-band, but all have pale-framed oval and kidney-shaped markings and a pale cross-line parallel to outer margin.

## STATUS AND COMMENTS
Widespread and generally common in central and S England, and in Wales; more local in N England; restricted to S lowlands in Scotland.

**KEY FACT** Like the Small Quaker, the Common Quaker is often abundant in early spring. After dark and with the aid of a torch, it can be seen feeding in good numbers at the catkins of sallows.

## SPOTTER'S CHART

| LOCATION | DATE/TIME |
|---|---|
| | |
| | |
| | |
| | |
| | |
| | |

# CLOUDED DRAB

*Orthosia incerta*

## FACT FILE

**SIZE** Length 18–22mm **HABITAT** Woodland, hedgerows and gardens **LARVAL FOODPLANT** Deciduous trees and shrubs, including oaks, sallows and birches **FLIGHT PERIOD** Mar–May

## IDENTIFICATION

Rests with its wings held in a shallow tent-like manner, forewings concealing hindwings. Forewing ground colour varies: typically yellow-buff, grey-buff, pinkish lilac or brown. Pale forms show a dark, wavy central band that encloses a pale-framed kidney-shaped mark, and a dark mark near tip of leading edge; these are harder to discern in dark forms.

## SPOTTER'S CHART

| LOCATION | DATE/TIME |
|---|---|
| | |
| | |
| | |
| | |
| | |
| | |

## KEY FACT

The Clouded Drab's common name belies its subtly beautiful appearance. Dark forms can be a struggle to identify, so get to know its commoner pale forms first to get a 'feel' and understanding of the species' character.

## STATUS AND COMMENTS

Widespread and generally common throughout much of England, Wales and Scotland, but least numerous in upland and N districts.

# HEBREW CHARACTER

*Orthosia gothica*

## FACT FILE

**SIZE** Length 20mm **HABITAT** Almost all habitats except the highest mountains **LARVAL FOODPLANT** Wide range of deciduous trees and shrubs **FLIGHT PERIOD** Mar–Apr

## IDENTIFICATION

Distinctive moth that rests with its wings held in a shallow tent-like manner, forewings concealing hindwings. Forewings are usually brown or grey-brown, marbled with darker and lighter areas; they also show a striking black mark, which resembles a saddle or an oblong that has had a semicircular hole removed by a hole-punch.

## SPOTTER'S CHART

| LOCATION | DATE/TIME |
| --- | --- |
| | |
| | |
| | |
| | |
| | |
| | |
| | |

## STATUS AND COMMENTS

Widespread and often extremely common throughout much of England, Wales and Scotland; least numerous in upland areas. One of the commonest early-spring moths.

## KEY FACT

The Hebrew Character has unique markings for a moth that flies in early spring, making it easy to recognise. The superficially similar Setaceous Hebrew Character is not on the wing until Jul.

# BROWN-LINE BRIGHT-EYE

*Mythimna conigera*

**FACT FILE**

**SIZE** Length 20–22mm **HABITAT** Grassland, waste ground and gardens **LARVAL FOODPLANT** Various grasses, including Cock's-foot and Common Couch **FLIGHT PERIOD** Jun–Jul

## IDENTIFICATION

Aptly named moth that rests with its wings spread flat or in a shallow tent-like manner, forewings concealing hindwings. Forewing tips are pointed and moth's overall outline is triangular. Forewings are buffish brown or reddish brown with a white teardrop-shaped central mark and dark brown cross-lines.

**KEY FACT** It is hard to confuse the Brown-line Bright-eye with any other moth because its forewing markings are so distinctive. It sometimes flies at dusk and is occasionally seen visiting flowers in the garden to feed.

## STATUS AND COMMENTS

Widespread and fairly common throughout much of England, Wales and Scotland, but least numerous in uplands and areas of intensive agriculture.

## SPOTTER'S CHART

| LOCATION | DATE/TIME |
|---|---|
| | |
| | |
| | |
| | |
| | |
| | |

# COMMON WAINSCOT

*Mythimna pallens*

## FACT FILE

**SIZE** Length 20mm **HABITAT** Grassland, waste ground and gardens **LARVAL FOODPLANT** Various grasses, including Common Couch Grass **FLIGHT PERIOD** Jun–Jul and Aug–Sep

## IDENTIFICATION

Rather plain-looking moth that usually rests with its wings held in a shallow tent-like manner, forewings concealing hindwings. Forewings are typically straw-coloured or yellow-buff, with whitish veins and faint, dark striations in between; there are small black dots alongside the main, central pale vein.

## STATUS AND COMMENTS

Widespread and locally common moth in much of central and S England, and in Wales; local in N England; restricted mainly to S lowlands in Scotland.

## KEY FACT

When resting, the Common Wainscot often aligns itself lengthways on a grass stalk or dried stem, where its markings allow it to blend in remarkably well. It is attracted to light and is often caught in moth traps.

## SPOTTER'S CHART

| LOCATION | DATE/TIME |
| --- | --- |
| | |
| | |
| | |
| | |
| | |

# BLACK RUSTIC

*Aporophyla nigra*

**FACT FILE**

**SIZE** Length 19–22mm **HABITAT** Grassland, moors, heaths, hedgerows and rural gardens **LARVAL FOODPLANT** Wide range of low-growing plants **FLIGHT PERIOD** Sep–Oct

## IDENTIFICATION

Distinctive moth that usually rests with its wings spread flat, or held in a shallow tent-like manner, forewings concealing hindwings. Forewings are relatively narrow and moth's outline is rather triangular overall. Forewings are dark sooty grey, marbled with black and with a striking white comma mark halfway along.

**KEY FACT** The Black Rustic has unique colours and markings for a moth of this size flying so late in the season. It is attracted to light and is caught in moth traps, and it sometimes visits garden flowers at dusk to feed.

## STATUS AND COMMENTS

Widespread and locally common in England and Wales, but most numerous in S England and least so in E England; local in Scotland.

## SPOTTER'S CHART

| LOCATION | DATE/TIME |
| --- | --- |
| | |
| | |
| | |
| | |

# EARLY GREY

*Xylocampa areola*

## FACT FILE

**SIZE** Length 17–19mm **HABITAT** Woodland, hedgerows and mature gardens **LARVAL FOODPLANT** Honeysuckle **FLIGHT PERIOD** Mar–May

## IDENTIFICATION

Well-marked moth that rests with its wings held in a shallow tent-like manner, forewings concealing hindwings. Forewings are typically grey or greyish brown overall, with darker marbling and markings, notably an irregular, lengthways dark line close to trailing edge and a row of dark streaks near margin.

## STATUS AND COMMENTS

Widespread and fairly common in S and central England, and in Wales; more local in N England; restricted mainly to lowland areas in Scotland.

## KEY FACT

There are superficially similar moths in terms of size, shape and markings, but the Early Grey is the most likely contender to be flying in early spring. It is attracted to light in small numbers.

## SPOTTER'S CHART

| LOCATION | DATE/TIME |
| --- | --- |
| | |
| | |
| | |
| | |
| | |

# GREEN-BRINDLED CRESCENT

*Allophyes oxyacanthae*

FACT FILE

**SIZE** Length 19–21mm **HABITAT** Woodland margins, hedgerows and gardens **LARVAL FOODPLANT** Wide range of deciduous trees and shrubs **FLIGHT PERIOD** Sep–Nov

## IDENTIFICATION

Beautifully patterned moth that rests with its wings held in a shallow tent-like manner, forewings concealing hindwings. Forewings are usually brown overall with green marbling, particularly along inner half of trailing edge; has a pale border to outer margin, pale buff kidney-shaped and rounded spots, and a pale thorax margin.

## KEY FACT

Confusingly, some forms of the Green-brindled Crescent lack green in the forewing colours; instead, the green, as well as the pale wing markings, are replaced by orange-brown. However, the overall pattern on the wings remains the same.

## STATUS AND COMMENTS

Widespread and locally common in much of England and Wales, although least numerous in the N; widespread in Scotland, but restricted mainly to lowland areas.

## SPOTTER'S CHART

| LOCATION | DATE/TIME |
|---|---|
| | |
| | |
| | |
| | |
| | |

# MERVEILLE DU JOUR

*Dichonia aprilina*

## FACT FILE

**SIZE** Length 20–22mm **HABITAT** Broadleaved woodland, hedgerows, parks and gardens **LARVAL FOODPLANT** Flowers, then leaves, of oaks **FLIGHT PERIOD** Sep–Oct

## IDENTIFICATION

Stunningly attractive moth that usually rests with its wings held in a shallow tent-like manner, forewings concealing hindwings. Overall outline is triangular. Forewings are an almost unique (among moths) shade of bluish green, with jagged black cross-lines and paler blue kidney-shaped and round spots.

## STATUS AND COMMENTS

Widespread and locally common in much of England and Wales, and probably most numerous in old woodlands in the S; widespread but local in Scotland.

## KEY FACT

The amazing colours and markings on the wings of the Merveille du Jour are a perfect match for certain species of lichen that grow on the branches and trunks of mature oaks.

## SPOTTER'S CHART

| LOCATION | DATE/TIME |
| --- | --- |
| | |
| | |
| | |
| | |
| | |
| | |

# THE SATELLITE

*Eupsilia transversa*

**FACT FILE**

**SIZE** Length 22–24mm **HABITAT** Woodland, hedgerows and gardens **LARVAL FOODPLANT** Deciduous trees and shrubs **FLIGHT PERIOD** Sep–Oct, then Mar–Apr after hibernation

## IDENTIFICATION

Distinctive moth that rests with its wings spread flat, one forewing largely overlapping the other and concealing hindwings. Forewings are a rich orange-brown with faint dark cross-lines and an oval to rounded spot; this can be white or orange, and has two tiny, similarly coloured satellite spots in 'orbit'.

## KEY FACT

Although The Satellite can sometimes be found flying on mild nights throughout the winter months, most individuals hibernate, amongst fallen leaves or in a tangle of dense Ivy.

## STATUS AND COMMENTS

Widespread and locally common moth in central and S England, and in Wales; more local in N England; confined mainly to S lowlands in Scotland.

## SPOTTER'S CHART

| LOCATION | DATE/TIME |
|---|---|
| | |
| | |
| | |
| | |
| | |
| | |
| | |

# THE CHESTNUT

*Conistra vaccinii*

## FACT FILE

**SIZE** Length 16–18mm **HABITAT** Broadleaved woodland, hedgerows, parks and gardens **LARVAL FOODPLANT** Deciduous trees and shrubs **FLIGHT PERIOD** Sep–May

## IDENTIFICATION

Richly coloured moth that usually rests with its wings spread flat, one forewing partly overlapping the other and concealing hindwings. Forewings have a curved leading edge and rather straight outer margin, and forewings are reddish brown with wavy blue-grey cross-lines and a dark grey central spot.

## SPOTTER'S CHART

| LOCATION | DATE/TIME |
| --- | --- |
| | |
| | |
| | |
| | |
| | |
| | |
| | |

## KEY FACT

Although The Chestnut hibernates and remains inactive during really cold spells, it flies on mild nights throughout winter. This means that it has one of the longest flight periods of any British moth.

## STATUS AND COMMENTS

Widespread and generally common moth throughout much of England, Wales and Scotland; perhaps least numerous in N and upland regions.

# RED-LINE QUAKER

*Agrochola lota*

FACT FILE

**SIZE** Length 17–19mm **HABITAT** Open woodland, hedgerows, heaths and gardens **LARVAL FOODPLANT** Various species of willow **FLIGHT PERIOD** Sep–Oct

## IDENTIFICATION

Well-marked moth that usually rests with its wings spread flat, forewings sometimes partly overlapping, and with hindwings concealed. Forewings are purplish grey with a rich red line parallel to outer margins, and red-framed oval and kidney-shaped markings, the latter containing a dark spot.

## SPOTTER'S CHART

| LOCATION | DATE/TIME |
| --- | --- |
| | |
| | |
| | |
| | |
| | |
| | |

## STATUS AND COMMENTS

Widespread and generally common in suitable habitats throughout most of England and Wales; least numerous in the N, and only local in Scotland.

## KEY FACT

The Red-line Quaker's larva hatches from its egg in early spring and feeds at first on willow catkins; later, once the catkins have finished, it feeds on young leaves. The adult is attracted to light.

# CENTRE-BARRED SALLOW

*Atethmia centrago*

## FACT FILE

**SIZE** Length 17–20mm **HABITAT** Woodland, hedgerows, parks and gardens **LARVAL FOODPLANT** Ash **FLIGHT PERIOD** Aug–Sep

## IDENTIFICATION

Colourful moth that rests with its wings spread flat, forewings concealing hindwings. Forewing outer margin is slightly scalloped. Forewings are yellow with a dark reddish-brown margin and central band; inner edge of latter is very straight. Head and thorax are reddish brown.

## KEY FACT

The larva of the Centre-barred Sallow hatches in spring and feeds on the buds of Ash; when these have opened, it then feeds on Ash flowers. It is mainly nocturnal in its habits.

## STATUS AND COMMENTS

Widespread and generally common in England and Wales, although its precise occurrence is dictated by the presence of its larval foodplant; local in Scotland, mainly in lowlands.

## SPOTTER'S CHART

| LOCATION | DATE/TIME |
|---|---|
| | |
| | |
| | |
| | |
| | |
| | |

# PINK-BARRED SALLOW

*Xanthia togata*

**FACT FILE**

**SIZE** Length 17–19mm **HABITAT** Damp woodland, wetland margins and damp heaths **LARVAL FOODPLANT** Sallow catkins at first, then herbaceous plants **FLIGHT PERIOD** Sep–Oct

## IDENTIFICATION

Stunningly beautiful moth that rests with its wings spread flat, or held in a shallow tent-like manner, forewings concealing hindwings. Wings are rich yellow, marked with reddish-brown spots, central band and narrow outer margin. Head is also flushed reddish brown.

**KEY FACT** The markings and colours of the Pink-barred Sallow are extremely striking when the moth is seen in isolation. But when it is resting on, or among, autumn leaves, it blends in remarkably well and is hard to spot.

## STATUS AND COMMENTS

Widespread and locally common moth throughout much of England and Wales in suitable damp habitats; widespread but more local in Scotland.

## SPOTTER'S CHART

| LOCATION | DATE/TIME |
| --- | --- |
| | |
| | |
| | |
| | |
| | |
| | |

# POPLAR GREY

*Acronicta megacephala*

## FACT FILE

**SIZE** Length 22–24mm **HABITAT** Open woodland, parks and gardens **LARVAL FOODPLANT** Mainly Aspen and poplars **FLIGHT PERIOD** Jun–Aug

## IDENTIFICATION

Usually rests with its wings spread flat or held in a shallow tent-like manner, forewings concealing hindwings. Outline is broadly triangular and forewing margins are rather rounded. Forewings are grey, subtly mottled with blackish grey and with black stippling and cross-lines. There is a pale, round central spot.

## STATUS AND COMMENTS

Widespread and fairly common in England and Wales, its precise occurrence dictated by the presence of its larval foodplants; widespread but local in Scotland.

## KEY FACT

The forewings have an unusual powdery appearance that allows the moth to blend in remarkably well when resting on the bark of large poplar and Aspen trees.

## SPOTTER'S CHART

| LOCATION | DATE/TIME |
|---|---|
| | |
| | |
| | |
| | |
| | |
| | |

# LUNAR UNDERWING

*Omphaloscelis lunosa*

**FACT FILE**

**SIZE** Length 18–23mm **HABITAT** Grassy habitats, including meadows, verges, parks and gardens **LARVAL FOODPLANT** Various grasses, including Yorkshire Fog **FLIGHT PERIOD** Aug–Oct

## IDENTIFICATION

Rests with its wings spread flat, one forewing partly overlapping the other and concealing hindwings. Ground colour of forewings is variable, but usually yellow-brown, reddish brown or grey-brown. Wings are marked with a network of pale veins and cross-lines and dark cross-bands; this pattern is fairly constant among individuals.

**KEY FACT** The Lunar Underwing gets its name from the dark markings, shaped like a crescent moon, seen on the otherwise pale hindwings. Although concealed at rest, the underwing is often exposed when the moth 'revs up' before taking flight.

## STATUS AND COMMENTS

Widespread and locally common moth across much of Wales and S and central England; local in N England; found mainly in S lowlands in Scotland.

## SPOTTER'S CHART

| LOCATION | DATE/TIME |
| --- | --- |
| | |
| | |
| | |
| | |
| | |
| | |

# STRAW UNDERWING

*Thalpophila matura*

## FACT FILE

**SIZE** Length 21–23mm **HABITAT** Grassland, moors and gardens **LARVAL FOODPLANT** Various grasses, including Mat-grass **FLIGHT PERIOD** Jul–Aug

## IDENTIFICATION

Well-marked moth that rests with its wings spread flat (on flat tree bark) or held in a tent-like manner (on a twig), forewings concealing hindwings. Forewings are grey-brown, marked and stippled with dark grey, and showing black and white cross-lines and an orange-brown eyespot.

## STATUS AND COMMENTS

Widespread and common in central and S Britain, but least so in SW England; local in N England; mainly coastal in Scotland.

## KEY FACT

The hindwings of this moth are a bright straw yellow. Although they are concealed when the moth is resting, the distinctively coloured hindwings are revealed as it spreads its forewings when agitated or about to take flight. There are no other common species that share this feature.

## SPOTTER'S CHART

| LOCATION | DATE/TIME |
|---|---|
| | |
| | |
| | |
| | |
| | |
| | |

# THE SYCAMORE
*Acronicta aceris*

FACT FILE

**SIZE** Length 22–26mm **HABITAT** Woodland, parks and gardens **LARVAL FOODPLANT** Mainly Horse-chestnut, but also Sycamore and Field Maple **FLIGHT PERIOD** Jun–Aug

## IDENTIFICATION
Subtly attractive moth that usually rests with its wings spread flat, forewings concealing hindwings. Outline is broadly triangular. Forewings are grey, stippled and marbled with dark grey and brown, and showing dark, jagged cross-lines and a dark-framed rounded spot.

## KEY FACT
The larva of The Sycamore is one of the most colourful and distinctive in Britain. It is covered in tufts of long yellow and orange hairs, and has a row of black-framed white spots on its dorsal surface.

## STATUS AND COMMENTS
Widespread and locally common in S Britain, with suburban SE England the main centre of its range; scarce in, or absent from, other parts of the region.

## SPOTTER'S CHART

| LOCATION | DATE/TIME |
| --- | --- |
| | |
| | |
| | |
| | |
| | |
| | |

# GREY DAGGER

*Acronicta psi*

## FACT FILE

**SIZE** Length 23–24mm **HABITAT** Woodland, hedgerows, parks and gardens **LARVAL FOODPLANT** Deciduous trees and shrubs **FLIGHT PERIOD** Jun–Aug

## IDENTIFICATION

Distinctively patterned moth that rests with its wings spread flat (on leaves or tree bark), or held in a tent-like manner (on twigs), forewings concealing hindwings. Forewings are grey and heavily stippled with dark grey, and show striking black dagger-like markings and fine, jagged cross-lines.

## STATUS AND COMMENTS

Widespread and locally common moth in much of central and S England, and in Wales; rather local in N England; restricted to lowlands in Scotland.

## KEY FACT

The Grey Dagger's larva is colourful and distinctive: the flanks are marked with red spots on each segment, the dorsal surface is yellow with a black margin, and there is a prominent black 'hump' at the head end.

## SPOTTER'S CHART

| LOCATION | DATE/TIME |
|---|---|
| | |
| | |
| | |
| | |
| | |
| | |

# KNOT GRASS

*Acronicta rumicis*

**FACT FILE**

**SIZE** Length 20–24mm **HABITAT** Wide range of open habitats, including parks, allotments and gardens **LARVAL FOODPLANT** Wide range of low-growing herbaceous plants **FLIGHT PERIOD** May–Jul

## IDENTIFICATION

Rests with its wings spread flat (on leaves or tree bark) or held in a tent-like manner (on twigs), forewings concealing hindwings. Forewings are variably grey, stippled and marbled with darker grey; has a jagged white band near outer margin, adjacent white marks midway along trailing edge, and a dark-centred pale eyespot.

## SPOTTER'S CHART

| LOCATION | DATE/TIME |
|---|---|
| | |
| | |
| | |
| | |
| | |
| | |
| | |

## STATUS AND COMMENTS

Widespread and locally common in much of central and S England, and in Wales; rather local in N England; restricted to lowlands in Scotland.

**KEY FACT**

Although the species is variable in appearance, its forewing markings usually allow identification. The eyespot, sited in a pale central cross-band, and the usually pale outer half of the trailing edge, look a bit like a pair of eyes and a nose.

# THE CORONET

*Craniophora ligustri*

## FACT FILE

**SIZE** Length 21–23mm **HABITAT** Open woodland, scrub, hedgerows, parks and gardens **LARVAL FOODPLANT** Mainly Ash and Wild Privet **FLIGHT PERIOD** Jun–Jul

## IDENTIFICATION

Beautifully patterned moth that usually rests with its wings spread flat, forewings concealing hindwings. Forewing ground colour is variable, but usually sooty-grey overall, marbled with brown and olive-green; has a pale cross-band, pale spot, and pale patch near outer margin that fancifully resembles a crown.

## KEY FACT

The striking markings on the forewings of The Coronet blend in remarkably well with lichens growing on tree bark. It is attracted to light after dark and is regularly caught in moth traps.

## STATUS AND COMMENTS

Widespread and common in S England and Wales, especially in the W; more local in central and N England; found mainly in the S and W in Scotland.

## SPOTTER'S CHART

| LOCATION | DATE/TIME |
|---|---|
| | |
| | |
| | |
| | |
| | |
| | |

# SVENSSON'S COPPER UNDERWING

*Amphipyra berbera*

**FACT FILE**

**SIZE** Length 26mm **HABITAT** Broadleaved woodland, hedgerows, parks and gardens **LARVAL FOODPLANT** Deciduous trees and shrubs, particularly Pedunculate Oak **FLIGHT PERIOD** Aug–Sep

## IDENTIFICATION

Usually rests with its wings spread flat, one forewing just overlapping the other and concealing both hindwings. Forewings are marbled brown, stippled with dark dots, and with jagged white cross-lines and a dark-centred pale spot. When the moth is agitated, coppery-orange underwings are sometimes revealed.

## STATUS AND COMMENTS

Widespread and common in S and central England, and in S Wales; more local in N England and elsewhere in Wales; status uncertain in Scotland.

## KEY FACT

Often rests in sheds and indoors during the daytime. The very similar Copper Underwing has less extensive orange on the underside of the hindwings, and pale brown palps (like tiny horns between the eyes; in Svensson's, palps are dark with white tips).

## SPOTTER'S CHART

| LOCATION | DATE/TIME |
|---|---|
| | |
| | |
| | |
| | |
| | |
| | |

# OLD LADY

*Mormo maura*



**FACT FILE**

**SIZE** Length 32mm **HABITAT** Woodland, hedgerows and gardens **LARVAL FOODPLANT** Herbaceous plants in autumn, shrubs in spring **FLIGHT PERIOD** Jul–Aug

## IDENTIFICATION

Broad-winged moth that rests with its wings spread flat, forewings concealing hindwings. Outline is triangular overall. Forewings are grey-brown with a broad, dark brown central band and subtle, jagged, dark cross-lines. Has a pale patch near tip of forewings, and head, thorax and base of forewings are dark brown.

## SPOTTER'S CHART

| LOCATION | DATE/TIME |
|---|---|
| | |
| | |
| | |
| | |
| | |
| | |

## STATUS AND COMMENTS

Widespread and fairly common in suitable damp habitats in S and central England, and in Wales; local further N in England; scarce in S Scotland.

**KEY FACT**

The Old Lady often hides indoors during the daytime, entering through an open window. The moth is easily disturbed and can be quite alarming when it first takes to the wing.

# ANGLE SHADES
*Phlogophora meticulosa*

FACT FILE

**SIZE** Length 27mm **HABITAT** Wide range of habitats, including parks and gardens **LARVAL FOODPLANT** Wide range of herbaceous and low-growing woody plants **FLIGHT PERIOD** May–Oct

## IDENTIFICATION
Unique and unmistakable moth that rests with its wings held flat, typically folded or creased lengthways; forewings conceal hindwings. Forewings are rather narrow and pointed, with jagged margins; they are typically marbled buff and brown, with a striking dark 'V' running from leading edge across middle.

### KEY FACT
The larva of the Angle Shades often feeds on the leaves of pot plants, including pelargoniums; these are usually kept indoors in the warmth in winter, which often causes the moths to emerge in midwinter.

## STATUS AND COMMENTS
Widespread and generally common in much of England and Wales, particularly in the S, where immigrants from Europe boost numbers; local in Scotland.

## SPOTTER'S CHART

| LOCATION | DATE/TIME |
|---|---|
| | |
| | |
| | |
| | |
| | |
| | |

# THE DUN-BAR

*Cosmia trapezina*

## FACT FILE

**SIZE** Length 17–21mm **HABITAT** Woodland, hedgerows, parks and gardens **LARVAL FOODPLANT** Deciduous trees and shrubs **FLIGHT PERIOD** Jul–Sep

## IDENTIFICATION

Rests with wings spread flat, forewings concealing hindwings. Forewing ground colour is variable, usually ranging from yellow-buff to grey-brown. Typical feature is a broad, central band that is usually darker than ground colour, defined by black and with a dark central dot; inner margin of band is sharply angled.

## KEY FACT

Most moth larvae are exclusively vegetarian, but not so The Dun-bar, whose tastes are more eclectic: it will eat the larvae of other moths, and will even engage in cannibalism if it encounters one of its own kind.

## SPOTTER'S CHART

| LOCATION | DATE/TIME |
| --- | --- |
| | |
| | |
| | |
| | |
| | |
| | |

## STATUS AND COMMENTS

Widespread and generally common throughout much of England and Wales; widespread but local in N England; found mainly in lowlands in Scotland.

# DARK ARCHES

*Apamea monoglypha*

FACT FILE

**SIZE** Length 23–29mm **HABITAT** Grassy habitats, including woods, hedgerows, parks and gardens **LARVAL FOODPLANT** Various grasses, including Cock's-foot **FLIGHT PERIOD** Jul–Aug

## IDENTIFICATION

Subtly attractive and well-marked moth that rests with its wings spread flat, or held in a shallow tent-like manner, forewings concealing hindwings. Forewings are marbled grey, stippled with dark dots, and with numerous dark streaks and a jagged black and white line near outer margin.

**KEY FACT** Although grey forms predominate in most parts, dark, melanic individuals of the Dark Arches are often found in N Britain. The species comes to light after dark and is often caught in moth traps in good numbers.

## STATUS AND COMMENTS

Widespread and often extremely common throughout much of England, Wales and Scotland, least so in upland areas. Can be extremely common in rural gardens.

## SPOTTER'S CHART

| LOCATION | DATE/TIME |
| --- | --- |
| | |
| | |
| | |
| | |
| | |
| | |

# CLOUDED BRINDLE

*Apamea epomidion*

**FACT FILE**

**SIZE** Length 21–24mm **HABITAT** Broadleaved woodland, mature hedgerows and rural gardens **LARVAL FOODPLANT** Various grasses, including Tufted Hair-grass **FLIGHT PERIOD** Jun–Jul

## IDENTIFICATION

Attractive moth that rests with its wings spread flat or held in a shallow tent-like manner, forewings concealing hindwings. Forewings are beautifully marbled buff and reddish brown, with a broad, dark central band containing a pale, rounded spot, and a dark half-band near outer margin; sides of the thorax are dark brown.

## STATUS AND COMMENTS

Widespread and fairly common in S and central England and Wales; rather local in N England; restricted mostly to S lowlands in Scotland.

**KEY FACT**

The markings and colours on the forewings of the Clouded Brindle are a perfect match for timber whose bark has fallen or been stripped away. Resting individuals are sometimes found by careful searching in the daytime.

## SPOTTER'S CHART

| LOCATION | DATE/TIME |
|---|---|
| | |
| | |
| | |
| | |
| | |
| | |

# MARBLED MINOR

*Oligia strigilis*

FACT FILE

**SIZE** Length 15–18mm **HABITAT** Grassy habitats, including woodland rides, parks and gardens **LARVAL FOODPLANT** Various grasses, including Common Couch **FLIGHT PERIOD** May–Jul

## IDENTIFICATION

Rather small, variable moth that rests with its wings spread flat, forewings concealing hindwings. Outline is rather triangular. Ground colour of forewings is variable, but often marbled grey-brown or rufous brown; typically, there is a pale submarginal band and a pale patch at base of trailing edge.

## KEY FACT

Several closely related and superficially very similar species are found in Britain; these cannot be identified reliably without dissection. Fortunately, the Marbled Minor is by far the commonest and most likely one to be encountered.

## SPOTTER'S CHART

| LOCATION | DATE/TIME |
| --- | --- |
| | |
| | |
| | |
| | |
| | |

## STATUS AND COMMENTS

Widespread and generally rather common throughout much of England, Wales and Scotland; found in almost all grassy habitats, except at high altitudes.

# COMMON RUSTIC
*Mesapamea secalis*

## FACT FILE
**SIZE** Length 16–20mm **HABITAT** Grassland, verges and gardens **LARVAL FOODPLANT** Various grasses, including Cock's-foot **FLIGHT PERIOD** Jul–Aug

## IDENTIFICATION
Rests with its wings spread flat, forewings concealing hindwings. Forewing ground colour varies from buffish brown, through reddish brown, to very dark, marbled brown. Typically, in paler specimens at least, a central dark band containing a pale kidney-shaped mark and, sometimes, a pale oval can be seen.

## STATUS AND COMMENTS
Widespread and generally common moth in central and S England, lowland Wales and N England; local in Scotland, restricted mainly to S lowlands.

## KEY FACT
The larva of the Common Rustic lives and feeds inside the stems of its foodplant grasses. The adult is attracted to light after dark and is regularly caught in moth traps.

## SPOTTER'S CHART

| LOCATION | DATE/TIME |
| --- | --- |
| | |
| | |
| | |
| | |
| | |

# FROSTED ORANGE

*Gortyna flavago*

FACT FILE

**SIZE** Length 20–23mm **HABITAT** Grassland, verges and waste ground **LARVAL FOODPLANT** Wide range of herbaceous plants, including thistles and burdocks **FLIGHT PERIOD** Aug–Oct

## IDENTIFICATION

Beautifully patterned moth that rests with its wings spread flat or held in a shallow tent-like manner, forewings concealing hindwings. Forewings are overall yellow-buff, stippled with orange-brown, and with pale round and kidney-shaped spots and a dark outer margin with a pale tip.

## KEY FACT

The larva of the Frosted Orange lives and feeds inside the stems of its foodplant. The patterns and markings on the wings are a good match for a yellowing, dying leaf.

## STATUS AND COMMENTS

Widespread and fairly common in central and S England, and in Wales; more local in N England; found mainly in lowland and coastal areas in Scotland.

## SPOTTER'S CHART

| LOCATION | DATE/TIME |
| --- | --- |
| | |
| | |
| | |
| | |

# TREBLE LINES

*Charanyca trigrammica*

**FACT FILE** **SIZE** Length 19–22mm **HABITAT** Rough open ground, including hedgerows, verges and gardens **LARVAL FOODPLANT** Wide range of herbaceous plants **FLIGHT PERIOD** May–Jul

## IDENTIFICATION

Usually rests with its wings spread flat, forewings concealing hindwings. Tips of forewings are rather rounded and ground colour is usually buffish brown; margin is typically subtly dark and there are three dark cross-lines. Head and thorax are furry and typically darker than forewing ground colour.

**KEY FACT** The larva of the Treble Lines is hard to find, as it lives at or below ground level and feeds only after dark. The adult is attracted to light after dark and is regularly caught in moth traps.

## STATUS AND COMMENTS

Widespread and generally common in S and central England, and in S Wales; local in N Wales and N England; absent from Scotland.

## SPOTTER'S CHART

| LOCATION | DATE/TIME |
|---|---|
| | |
| | |
| | |
| | |
| | |
| | |

# THE HERALD
*Scoliopteryx libratrix*

**FACT FILE**

**SIZE** Length 20mm **HABITAT** Broadleaved woodland, hedgerows, parks and gardens **LARVAL FOODPLANT** Sallows and willows **FLIGHT PERIOD** Sep–Nov, and Mar–Jun after hibernation

## IDENTIFICATION
Unmistakable moth that rests with its wings spread flat or held in a low, curved arch, forewings concealing hindwings. Forewings have a distinctive jagged margin, and are brown with stippled orange patches in centre and near base, and striking white cross-lines, and white veins between central cross-line and outer margin.

## STATUS AND COMMENTS
Widespread and locally common in England and Wales, although its precise distribution is often patchy; local in the N; rather scarce in Scotland, where it is found mainly in the S.

## SPOTTER'S CHART

| LOCATION | DATE/TIME |
| --- | --- |
| | |
| | |
| | |
| | |
| | |

**KEY FACT**

The Herald hibernates during the winter months and is sometimes discovered resting on the inside wall of a garden shed or outhouse, and occasionally gathers in small groups in suitable locations.

# NUT-TREE TUSSOCK
*Colocasia coryli*

## FACT FILE

**SIZE** Length 18mm **HABITAT** Woodland, hedgerows and gardens **LARVAL FOODPLANT** Hazel and other deciduous trees **FLIGHT PERIOD** Apr–Jun and Jul-Sep in S. May–Jun in N

## IDENTIFICATION

Well-marked, furry-looking moth that rests with its wings held in a tent-like manner, forewings concealing hindwings. Forewings are greyish on outer half, marbled with brown and with jagged cross-lines; inner half is darker brown and contains a bold, dark-centred eyespot. Head, thorax and legs are greyish and furry.

## STATUS AND COMMENTS

Widespread and generally common moth in S and central England and Wales; more locally and patchily distributed in N England and Scotland.

## KEY FACT

The Nut-tree Tussock's markings are a very good match for tree bark; its camouflage is particularly effective when it rests on a twig with its wings partly wrapped around itself.

## SPOTTER'S CHART

| LOCATION | DATE/TIME |
| --- | --- |
| | |
| | |
| | |
| | |

# SILVER Y

*Autographa gamma*

FACT FILE

**SIZE** Length 21mm **HABITAT** Wide range of open habitats, including grassland, coasts and gardens **LARVAL FOODPLANT** Herbaceous plants **FLIGHT PERIOD** Overlapping broods in May–Oct

## IDENTIFICATION

Well-marked and almost unmistakable moth that rests with its wings held in a tent-like manner, forewings concealing hindwings. Forewings are beautifully marbled with grey and brown, and marked with a bold white 'Y' that gives the species its name. Note its humpbacked appearance in profile.

## KEY FACT

The Silver Y is a fast-flying moth whose status in Britain is mainly that of an annual immigrant from Europe, in large numbers. The likelihood is that it does not survive the winter in Britain.

## STATUS AND COMMENTS

Widespread and common throughout much of England and Wales, and particularly numerous near coasts; widespread but more local in Scotland.

## SPOTTER'S CHART

| LOCATION | DATE/TIME |
| --- | --- |
| | |
| | |
| | |
| | |
| | |
| | |

# BEAUTIFUL GOLDEN Y
*Autographa pulchrina*



**FACT FILE** **SIZE** Length 22–23mm **HABITAT** Woodland margins, hedgerows, waste ground and gardens **LARVAL FOODPLANT** Wide range of low-growing herbaceous plants **FLIGHT PERIOD** Jun–Jul

## IDENTIFICATION
Richly marked moth that rests with its wings held in a tent-like manner, forewings concealing hindwings. Forewings are reddish brown and marbled with pinkish buff; central white 'V' marking and adjacent white oval align to create the impression of a 'Y' marking. Hair tufts on thorax are obvious in profile.

## STATUS AND COMMENTS
Widespread and generally quite common moth in S and central England, and in Wales; more local in N England; mainly in lowlands in Scotland.

**KEY FACT**
With its marbled forewings, a freshly emerged Beautiful Golden Y is a truly stunning moth. It is attracted to light after dark in small numbers and is regularly caught in moth traps.

## SPOTTER'S CHART

| LOCATION | DATE/TIME |
| --- | --- |
| | |
| | |
| | |
| | |
| | |

# GREEN SILVER-LINES

*Pseudoips prasinana*

FACT FILE

**SIZE** Length 17mm **HABITAT** Broadleaved woodland, mature hedgerows and rural gardens **LARVAL FOODPLANT** Broadleaved trees, notably oaks and birches **FLIGHT PERIOD** Jun–Jul

## IDENTIFICATION

Subtly colourful moth that rests with its wings held in a tent-like manner, forewings concealing hindwings. Forewings green with three diagonal white cross-lines; outer margin is pinkish in male but whitish in female. Head and thorax are green and furry. Antennae and front legs are pinkish.

## KEY FACT

When the Green Silver-lines rests on the underside of a leaf, its colours and markings allow it to blend in remarkably well and avoid detection by would-be predators such as birds.

## SPOTTER'S CHART

| LOCATION | DATE/TIME |
|---|---|
| | |
| | |
| | |
| | |
| | |
| | |

## STATUS AND COMMENTS

Widespread and locally common across much of Wales and S and central England; more local in N England; confined mainly to S lowlands in Scotland.

# BURNISHED BRASS

*Diachrisia chrysitis*

## FACT FILE

**SIZE** Length 21mm **HABITAT** Hedgerows and gardens **LARVAL FOODPLANT** Mainly Common Nettle **FLIGHT PERIOD** Mostly Jun–Jul, sometimes with a 2nd brood in Aug

## IDENTIFICATION

Stunning and unmistakable moth that rests with its wings held in a tent-like manner, forewings concealing hindwings. Forewings are pinkish brown with large areas of metallic-looking, burnished golden yellow. Note the humpbacked appearance in profile, with striking tufts of hairs on thorax.

## STATUS AND COMMENTS

Widespread and generally common across much of England and Wales, least so in upland regions; commonest in lowland areas in Scotland.

## KEY FACT

Although the wing markings and colours of the Burnished Brass are striking when seen in isolation, the moth is remarkably well camouflaged when resting in dappled shade underneath a leaf or in leaf-litter.

## SPOTTER'S CHART

| LOCATION | DATE/TIME |
| --- | --- |
| | |
| | |
| | |
| | |
| | |
| | |

# GOLD SPOT

*Plusia festucae*

FACT FILE

**SIZE** Length 22mm **HABITAT** Fens, marshes and river margins **LARVAL FOODPLANT** Various sedges, Yellow Iris, and other plants of damp ground **FLIGHT PERIOD** Jun–Sep

## IDENTIFICATION

Striking moth that rests with its wings held in a tent-like manner, forewings concealing hindwings. Forewings are brown overall, with patches of yellowish between veins, two central white spots and a white streak near outer margin of leading edge. Thorax has prominent hair tufts, best seen in profile.

## STATUS AND COMMENTS

Widespread and very locally common moth in suitable habitats in S and central England and Wales; more local in N England and Scotland.

KEY FACT

The Gold Spot can sometimes be seen on the wing in the early evening, visiting marshland flowers to feed. It is attracted to light in small numbers and is sometimes caught in moth traps.

## SPOTTER'S CHART

| LOCATION | DATE/TIME |
| --- | --- |
| | |
| | |
| | |
| | |

# THE SNOUT

*Hypena proboscidalis*

**FACT FILE** SIZE Length 19–22mm HABITAT Woodland margins, hedgerows, waste ground and gardens LARVAL FOODPLANT Common Nettle FLIGHT PERIOD Jun–Aug

## IDENTIFICATION

Distinctive moth that rests with its wings spread flat, forewings concealing hindwings. Outline is distinctly triangular, with hook-tipped forewings; projecting palps form a 'snout'. Forewings are buffish brown to grey-brown with darker cross-lines, the outer two often enclosing and defining a subtly paler band.

## SPOTTER'S CHART

| LOCATION | DATE/TIME |
|---|---|
| | |
| | |
| | |
| | |
| | |
| | |

## KEY FACT

The Snout is easily disturbed from patches of its larval foodplant, Common Nettle, during the daytime. It often takes to the wing in the early evening and is regularly caught in moth traps.

## STATUS AND COMMENTS

Widespread and often locally very common across most of England and Wales, in places where the larval foodplant thrives; widespread but more local in Scotland.

# THE SPECTACLE
*Abrostola tripartita*

**FACT FILE**

**SIZE** Length 18mm **HABITAT** Rough grassland, waste ground, hedgerows and gardens **LARVAL FOODPLANT** Common Nettle **FLIGHT PERIOD** May–Jul

**KEY FACT** The Spectacle typically rests lengthways along a twig, wrapping its wingtips around the twig if its diameter is small enough. The species is attracted to light after dark and is often caught in moth traps.

## IDENTIFICATION
Distinctive moth that rests with its wings held in a tent-like manner or slightly rolled lengthways, forewings concealing hindwings. Forewings are grey overall, but with a dark central band defined by orange-buff lines. Has forward-projecting hair tufts on the thorax. Seen head-on, has white 'spectacle' markings on thorax.

## STATUS AND COMMENTS
Widespread and often locally very common across most of England and Wales, in places where the larval foodplant thrives; more local in Scotland, where it occurs mainly in lowlands.

## SPOTTER'S CHART

| LOCATION | DATE/TIME |
|---|---|
| | |
| | |
| | |
| | |
| | |
| | |

# RED UNDERWING

*Catocala nupta*

## FACT FILE

**SIZE** Wingspan 65mm **HABITAT** Broadleaved woodland, hedgerows, parks and gardens **LARVAL FOODPLANT** Willows and poplars **FLIGHT PERIOD** Aug–Sep

## IDENTIFICATION

Large and unmistakable moth that rests with its wings spread flat; forewings conceal hindwings if moth remains undisturbed, but these are revealed when it is alarmed. Forewings are beautifully marbled grey and brown, with jagged cross-lines. Hindwings are banded with black and red.

## STATUS AND COMMENTS

Widespread and locally common in central and S Britain, particularly in S and SE England; local in N England and N Wales; absent from Scotland.

## SPOTTER'S CHART

| LOCATION | DATE/TIME |
|---|---|
| | |
| | |
| | |
| | |
| | |
| | |

## KEY FACT

The species' forewing markings and colours are a good match for tree bark and provide excellent camouflage. If the moth is discovered by a predator, however, its red underwings are revealed to startle the would-be attacker.

# BEAUTIFUL HOOK-TIP

*Laspeyria flexula*

**FACT FILE**

SIZE **Wingspan 23–26mm** HABITAT **Broadleaved and coniferous woodland, parks and gardens** LARVAL FOODPLANT **Lichens growing on tree branches** FLIGHT PERIOD **Jun–Aug**

## IDENTIFICATION

A distinctive moth that rests with its wings spread flat, forewings concealing hindwings. Forewings have a wavy outer margin and hooked tip; they are buffish brown or greyish brown, stippled with dark dots and marked with two reddish-framed yellow cross-lines; tips are flushed deep reddish brown.

## SPOTTER'S CHART

| LOCATION | DATE/TIME |
|---|---|
| | |
| | |
| | |
| | |
| | |
| | |

**KEY FACT**

The markings and colours on the wings of the Beautiful Hook-tip are a good match for tree bark or cut timber. The moth is attracted to light and is caught in moth traps in small numbers.

## STATUS AND COMMENTS

Widespread and fairly common in S England and S Wales; more local in central England, and extending N to the Yorkshire border.

# STRAW DOT
*Rivula sericealis*

**FACT FILE** SIZE Length 15mm HABITAT Grassland, hedgerows and gardens LARVAL FOODPLANT Various grasses FLIGHT PERIOD Jun–Jul, and sometimes Aug–Sep as a 2nd brood

## IDENTIFICATION
Unmistakable little moth that rests with its wings spread flat, forewings concealing hindwings. Outline is broadly triangular and wingtips are pointed. Forewings are buffish yellow with a bold purplish-brown kidney-shaped spot and a reddish-buff outer margin. Worn specimens can look very pale but dark spot is usually still obvious.

## STATUS AND COMMENTS
Widespread and common in S and central England, and in Wales; more local in N England and Scotland, where it is found mainly in lowlands and coastal districts.

## KEY FACT
The Straw Dot is a resident species in Britain, but each year its numbers are boosted by influxes of immigrants from mainland Europe. It is attracted to light and is caught in moth traps.

## SPOTTER'S CHART

| LOCATION | DATE/TIME |
| --- | --- |
| | |
| | |
| | |
| | |
| | |

# INDEX

Common names are in plain text and scientific names are in *italic*.

# INDEX



## PHOTOGRAPHIC ACKNOWLEDGEMENTS

Photographs supplied by Nature Photographers Ltd. All photographs by Paul Sterry except for the those on the following pages:

Robin Bush: 8, 12, 25, 38; Andrew Merrick: 78, 98, 174; Richard Revels: 16, 27, 36, 43, 44, 59, 60, 64, 107.